# Methods and Protocols in Food Science

*Series Editor*
**Anderson S. Sant'Ana**
**University of Campinas**
**Campinas, Brazil**

For further volumes:
http://www.springer.com/series/16556

*Methods and Protocols in Food Science* series is devoted to the publication of research protocols and methodologies in all fields of food science.

Volumes and chapters will be organized by field and presented in such way that the readers will be able to reproduce the experiments in a step-by-step style. Each protocol will be characterized by a brief introductory section, followed by a short aims section, in which the precise purpose of the protocol will be clarified.

# Heat Resistant Microorganisms in Food Safety and Spoilage

Edited by

**Anderson S. Sant'Ana**

*Faculty of Food Engineering, University of Campinas, Campinas, Brazil*

Humana Press

*Editor*
Anderson S. Sant'Ana
Faculty of Food Engineering
University of Campinas
Campinas, Brazil

ISSN 2662-950X ISSN 2662-9518 (electronic)
Methods and Protocols in Food Science
ISBN 978-1-0716-5279-4 ISBN 978-1-0716-5280-0 (eBook)
https://doi.org/10.1007/978-1-0716-5280-0



This Humana imprint is published by the registered company Springer Science+Business Media, LLC, part of Springer Nature.
The registered company address is: 1 New York Plaza, New York, NY 10004, U.S.A.

## Preface to the Series

The Methods and Protocols in Food Science series is devoted to the publication of research protocols and methodologies in all fields of food science. The series is unique as it includes protocols developed, validated, and used by food and related scientists as well as provides theoretical basis for each protocol. Aspects related to improvements in the protocols and adaptations and further developments in the protocols may also be approached.

The Methods and Protocols in Food Science series aims to bring the most recent developments in research protocols in the field as well as very well-established methods. As such the series targets undergraduates, graduates, and researchers in the field of food science and correlated areas. The protocols documented in the series will be highly useful for scientific inquiries in the field of food sciences and are presented in such a way that the readers will be able to reproduce the experiments in a step-by-step style.

Each protocol will be characterized by a brief introductory section, followed by a short aims section, in which the precise purpose of the protocol is clarified. Then, an in-depth list of materials and reagents required for employing the protocol is presented, followed by comprehensive and step-by-step procedures on how to perform that experiment. The next section brings the dos and don'ts when carrying out the protocol, followed by the main pitfalls faced and how to troubleshoot them. Finally, template results will be presented and their meaning/conclusions addressed.

The Methods and Protocols in Food Science series will fill an important gap, addressing a common complaint of food scientists, regarding the difficulties in repeating experiments detailed in scientific papers. With this, the series has a potential to become a reference material in food science laboratories of research centers and universities throughout the world.

*Campinas, Brazil* *Anderson S. Sant'Ana*

# Preface

Thermal processing remains one of the most widely applied and reliable preservation strategies in the food industry. Despite its long history and extensive industrial use, the effective control of microorganisms through heat treatment continues to pose significant scientific and technological challenges. Among these, the presence of heat-resistant microorganisms—including bacterial spores, thermophilic bacteria, and heat-resistant molds—represents a critical concern for food safety, spoilage prevention, and process validation.

Protocols on Heat-Resistant Microorganisms was conceived to address the persistent gap between theoretical concepts of thermal inactivation and their practical implementation in laboratory and industrial settings. While thermal death kinetics, D- and Z-values, and process lethality are well established in the literature, reproducible and standardized experimental approaches for studying heat resistance remain difficult to implement, particularly when accounting for the complex effects of food matrices, processing equipment, and microbial physiological states.

This volume brings together a comprehensive collection of protocols that cover the preparation, handling, and thermal treatment of heat-resistant microorganisms, as well as the methods required for their detection, enumeration, and kinetic analysis. The chapters guide the reader through critical steps such as the production and standardization of spore suspensions, calibration and placement of temperature sensors, determination of heat penetration and cold spots, and the execution of commercial sterility and validation studies. Special attention is given to both classical log-linear models and non-log-linear inactivation behaviors, reflecting the growing recognition of shoulders, tailing, and other deviations from first-order kinetics in real food systems.

In addition to bacterial spores, this book addresses the often-overlooked role of heat-resistant molds and yeasts, which are increasingly relevant in the spoilage of acid and high-value food products. The inclusion of protocols related to *Alicyclobacillus*, *Geobacillus*, *Bacillus*, and *Clostridium* species reflects their industrial and regulatory importance, while the discussion of biological indicators and surrogate microorganisms supports the design of safe and robust thermal processes.

This book is intended for researchers, graduate students, food microbiologists, process authorities, and professionals involved in food processing, safety assessment, and regulatory compliance. By emphasizing practical guidance, critical control points, and methodological rigor, Protocols on Heat-Resistant Microorganisms aims to serve both as a laboratory manual and as a reference for designing, validating, and interpreting thermal processing studies under realistic conditions.

We hope that this volume will contribute to improving reproducibility, strengthening the scientific basis of thermal process design, and ultimately supporting the production of safe, stable, and high-quality foods in an evolving global food system.

*Campinas, São Paulo, Brazil* *Anderson S. Sant'Ana*

# Contents

## Contributors

DIONISIO PEDRO AMORIM NETO • *Department of Food Science and Nutrition, Faculty of Food Engineering, University of Campinas, Campinas, SP, Brazil*
JAQUELINE SOUSA CORREIA • *Department of Food Science and Nutrition, Faculty of Food Engineering, University of Campinas, Campinas, SP, Brazil*
EUGÉNIO DA PIEDADE EDMUNDO SITOE • *Department of Food Science and Nutrition, Faculty of Food Engineering, University of Campinas, Campinas, SP, Brazil*
ALESSANDRA REGINA DA SILVA • *Labtermo, Analysis, Research and Microbiological Consulting, Campinas, Brazil*
LEONARDO DO PRADO-SILVA • *Department of Clinical Analyses, Toxicology and Food Science, School of Pharmaceutical Sciences of Ribeirão Preto, University of São Paulo, São Paulo, Brazil*
LUÍSA FREIRE • *Faculty of Pharmaceutical Sciences, Food and Nutrition, Federal University of Mato Grosso do Sul, Campo Grande, Mato Grosso do Sul, Brazil*
WILMA CUSTÓDIO FUMO • *Department of Food Science and Nutrition, Faculty of Food Engineering, State University of Campinas, Campinas, SP, Brazil*
IZAEL GRESSONI JUNIOR • *Labtermo, Analysis, Research and Microbiological Consulting, Campinas, Brazil; UNICAMP – University of Campinas, Campinas, Brazil*
CAROLINE HECKLER • *Department of Food Science and Nutrition, Faculty of Food Engineering, University of Campinas, Campinas, SP, Brazil*
MATHEUS PÉRICLES SILVA LÁSCARIS • *Department of Food Science and Nutrition, Faculty of Food Engineering, University of Campinas, Campinas, SP, Brazil*
BEATRIZ DAL PIAN MACHADO • *Department of Food Science and Nutrition, Faculty of Food Engineering, University of Campinas, Campinas, SP, Brazil*
LARISSA PEREIRA MARGALHO • *Department of Food Science and Nutrition, Faculty of Food Engineering, University of Campinas, Campinas, SP, Brazil*
MATHEUS S. MOURÃO • *Department of Food Science and Nutrition, Faculty of Food Engineering, University of Campinas, Campinas, SP, Brazil*
DIONISIO PEDRO AMORIM NETO • *Department of Food Science and Nutrition, Faculty of Food Engineering, University of Campinas, Campinas, SP, Brazil*
ROBERTO PESSANHA DA SILVA PIRES • *Department of Food Science and Nutrition, Faculty of Food Engineering, University of Campinas, Campinas, SP, Brazil; Federal Institute of Education, Science and Technology of Rio de Janeiro, Rio de Janeiro, RJ, Brazil*
MARCELO FELIPE SILVA ESTÁCIO SANTANA • *Department of Food Science and Nutrition, Faculty of Food Engineering, State University of Campinas, Campinas, SP, Brazil*
MAHMOUD YOLMEH • *Department of Food Science and Nutrition, Faculty of Food Engineering, University of Campinas, Campinas, SP, Brazil*

# Chapter 1

# Enumeration of Heat-Resistant Microorganisms: Bacteria

**Caroline Heckler**


## Abstract

Heat-resistant bacteria, particularly spore-forming mesophilic and thermophilic bacteria, are critical to food safety and quality assurance in the food industry. These microorganisms are capable of surviving heat and chemical treatments due to their ability to form highly resistant spores, which can cause spoilage and pose potential risks if not adequately controlled. This chapter provides detailed protocols for enumerating four categories of spore-forming bacteria: mesophilic aerobic, mesophilic anaerobic, thermophilic aerobic, and thermophilic anaerobic.


**Key words** Bacterial spores, Quantitative microbiology, Spore enumeration, Microbiological analysis, Food safety

## 1 Introduction

Bacterial spore-formers are common contaminants of soil, water, air, and raw materials used in the manufacturing of food products [1, 2]. These organisms, belonging to different taxonomic groups, have the ability to form spores resistant to adverse conditions such as heat, extreme pH, and dehydration. This resilience allows them to persist in hostile environments, including rigorous industrial processes [2]. The main groups of spore-forming microorganisms of interest are aerobic mesophiles, anaerobic mesophiles, aerobic thermophiles, and anaerobic thermophiles.

Mesophilic aerobic spore-formers include genera such as *Bacillus*, *Paenibacillus*, *Brevibacillus*, and *Sporolactobacillus*. These microorganisms are frequently found in products like canned vegetables, dairy products, ingredients, and processed foods, and they can cause spoilage [3]. *B. cereus*, for example, is a relevant species in this group due to its role in food spoilage and its ability to produce toxins that cause foodborne illnesses such as diarrhea and vomiting [4]. Moreover, certain species of *Bacillus* and related microorganisms are responsible for rope spoilage in bread, characterized by sticky texture, the formation of threads in the crumb, a sweet odor

Anderson S. Sant'Ana (ed.), *Heat Resistant Microorganisms in Food Safety and Spoilage*, Methods and Protocols in Food Science, https://doi.org/10.1007/978-1-0716-5280-0_1,

reminiscent of ripe fruit, and, in advanced cases, liquefaction [5, 6]. This issue occurs due to the survival of heat-resistant spores, and their contamination is inevitable as they are part of the natural microbiota of wheat [5].

Mesophilic anaerobic spore-formers, such as those in the genus *Clostridium*, are divided into two main groups of importance in food: proteolytic (i.e., putrefactive) and nonproteolytic (i.e., saccharolytic). These microorganisms contaminate foods like meat, dairy products, and canned vegetables [7]. Proteolytic strains include *C. sporogenes*, proteolytic strains of *C. botulinum*, and other heat-resistant putrefactive anaerobes, causing unpleasant odors and gas production due to protein degradation [4, 7]. Nonproteolytic strains include *C. perfringens*, nonproteolytic strains of *C. botulinum* (type E), and butyric acid-producing anaerobes like *C. butyricum*. These microorganisms produce gas and organic acids, compromising the pH and texture of foods [4, 7]. In addition to spoilage, *C. botulinum* and *C. perfringens* are significant public health concerns due to their production of potent toxins associated with botulism and foodborne intoxications [4, 7].

Thermophilic aerobic spore-formers are notable for causing spoilage in canned and commercially sterilized foods under inadequate storage conditions at elevated temperatures [8]. Some microorganisms in this group, like *Heyndrickxia coagulans* and *Geobacillus stearothermophilus*, ferment carbohydrates, producing short-chain fatty acids such as lactic, acetic, and propionic acids, which give the product an acidic flavor. However, they do not produce enough gas to alter the flat appearance of the container ends, a spoilage phenomenon known as flat sour [8]. Another relevant microorganism in this group is *Alicyclobacillus* spp., which is a significant concern in acidic fruit juices, concentrates, and mixed juice products. Its thermoacidophilic nature and highly resistant spores allow its survival during production, resulting in spoilage characterized by chemical off-flavors [9].

Thermophilic anaerobic spore-formers can be divided into $H_2S$-producing and non-$H_2S$-producing types. Among non-$H_2S$ producers, *Thermoanaerobacterium thermosaccharolyticum* stands out for causing spoilage in inadequately processed canned products such as foods with tomato sauce, vegetables, mushrooms, soup, and dog food [10]. These organisms produce large amounts of gas and organic acids, leading to loss of vacuum in cans and pH reduction, compromising the stability of the products [10]. Among $H_2S$ producers, *Desulfotomaculum nigrificans* is notable for causing spoilage in preserved vegetables. This type of spoilage does not cause can swelling but produces a distinct hydrogen sulfide odor and blackening due to the reaction between sulfide and iron in the container [11].

Given the economic, food quality, and food safety impacts associated with these groups, precise enumeration of spores is essential for monitoring the quality of food products and ensuring the validation of industrial processes. This chapter describes specific procedures for the detection and quantification of spores of these microorganisms, considering their physiological and ecological characteristics based on international references [3, 7, 8, 10–17].

## 2 Materials

### 2.1 Enumeration of Mesophilic Aerobic Spores

1. Culture media and diluent:
   (a) Sterile 0.1% peptone water: Weigh 1.0 g of peptone and dissolve in 1 L of distilled water. Adjust the pH to 7.0 ± 0.1. Dispense the content in bottles. Autoclave the solution at 121 °C for 15 min.
   (b) Sterile tryptone glucose extract agar (TGEA): Use a commercial medium and follow the preparation and sterilization instructions provided by the manufacturer.
   (c) Sterile dextrose tryptone agar (DTA): Use a commercial medium and follow the preparation and sterilization instructions provided by the manufacturer.
2. Equipment:
   (a) Stomacher or an equivalent device for sample agitation (only for use with bags).
   (b) Incubator capable of maintaining a uniform temperature of 35 °C.
   (c) Thermostatically controlled water or oil baths set at 45 °C (tempering agar), 80 °C, and boiling (94–100 °C).
   (d) Colony counter.
3. Other tools:
   (a) Sterile bags or containers with a capacity ≥1 L, leakproof.
   (b) Sterile Erlenmeyer flasks, 500 mL or 250 mL (used only for rope-producing spores).
   (c) Sterile pipettes with capacities of 10 mL, 1 mL, and 0.1 mL, along with a pipette bulb for liquid handling.
   (d) Sterile Petri plates (90 × 15 mm).
   (e) Thermometer.

### 2.2 Enumeration of Mesophilic Anaerobic Spores

1. Culture media and diluent:
   (a) Sterile distilled water or sterile 0.1% peptone water (*see* **Item 2.1(a)**).
   (b) Sterile trypticase peptone glucose yeast extract (TPGY) agar: Weigh 50 g of trypticase or tryptone, 20 g of yeast extract, 15 g of bacteriological agar, 5 g of

peptone, 4 g of dextrose/glucose, and 1 g of sodium thioglycollate and dissolve in 1 L of distilled water. Dispense the content on bottles. Autoclave at 121 °C for 15 min. The final pH should be 7.0 ± 0.1. Cool and maintain it at 45 °C until use.

(c) Sterile tryptose cycloserine dextrose (TCD) agar: Weigh 15 g of tryptose, 5 g of soytone, 5 g of yeast extract, 1 g of ferric ammonium citrate, and 20 g of agar and dissolve in 1 L of distilled water. Adjust the pH to 7.6 ± 0.2. Autoclave the medium at 121 °C for 15 min and then cool it to approximately 50 °C in a water bath. Once the medium has cooled, add 10.0 mL of a filter-sterilized 4.0% D-cycloserine solution per liter of medium to achieve a final concentration of 400 mg/mL.

(d) Sterile anaerobic clostridial (AC) agar: Use a commercial medium and follow the preparation and sterilization instructions provided by the manufacturer.

(e) Sterile trypticase peptone (TP) agar: Weigh 50 g of trypticase, 5 g of peptone, 1 g of sodium thioglycollate, and 15 g of agar and dissolve in 1 L of distilled water. Sterilize at 121 °C for 15 min. After sterilization, add 14 mL of filter-sterilized 10% sodium bicarbonate to the tempered agar just before pouring it into Petri plates.

(f) Sterile thioglycolate agar: Weigh 29.5 g of dehydrated fluid thioglycollate medium and 20 g of agar and suspend in 1.0 L of distilled water. Heat to boiling [7] while stirring until completely dissolved. Distribute the medium into flasks and autoclave at 121 °C for 15 min. Use the medium within 1 week of preparation.

(g) Sterile and de-aerated PE-2 medium: Weigh 20 g of peptone, 3 g of yeast extract, and 2 mL of a 2% alcoholic bromocresol purple solution and dissolve in 1 L of water, applying gentle heat if needed. Pour 19 mL of the prepared solution into 20 × 150 mm screw-cap test tubes, each containing 8–10 untreated seed peas. Sterilize the tubes in an autoclave at 121 °C for 15 min. If the medium is not freshly prepared, heat it to boiling (*see* **Note** 7) for 20 min to remove oxygen and allow it to cool to 55 °C before use.

(h) Sterile and de-aerated cooked meat medium (CMM): Use a commercial medium and follow the preparation and sterilization instructions provided by the manufacturer. If the medium is not freshly prepared, heat it to boiling (*see* **Note** 7) for 20 min to remove oxygen and allow it to cool to 55 °C before use.

(i) Sterile petroleum jelly (also known as vaseline or petrolatum): Place petroleum jelly in a heat-resistant bottle. Place it in an oven and sterilize it at 191 °C for 3 h. Allow it to cool to 55 °C before use. If not used immediately, store the bottle in a clean, dry place, and reheat it in a water bath (60–70 °C) to liquefy before use.

(j) Sterile Vaspar: Combine one part mineral oil with two parts petroleum jelly in a heat-resistant bottle. Mix thoroughly until homogeneous. Place the bottle in an oven and sterilize it at 191 °C for 3 h. Allow it to cool to 55 °C before use. If not used immediately, store the bottle in a clean, dry place, and reheat it in a water bath (60–70 °C) to liquefy before use.

2. Equipment:
   (a) Stomacher or an equivalent device for sample agitation (only for use with bags).
   (b) Incubator capable of maintaining a uniform temperature of 30–35 °C.
   (c) Thermostatically controlled water or oil baths set at 45 °C (tempering agar), 60 °C, 71 °C, or 80 °C.
   (d) Colony counter.
3. Other tools:
   (a) Sterile bags or containers with a capacity ≥500 mL, leakproof.
   (b) Water and ice bath.
   (c) Sterile pipettes with capacities of 1 mL and 0.1 mL, along with a pipette bulb for liquid handling.
   (d) Sterile Petri plates (90 × 15 mm).
   (e) Sterile, empty tubes (16 × 125 mm and 20 × 150 mm).
   (f) Anaerobic jar with an anaerobic atmosphere generation system or an anaerobic chamber.
   (g) Thermometer.

### 2.3 Enumeration of Thermophilic Aerobic Spores

1. Culture media and diluent:
   (a) Sterile distilled water.
   (b) Sterile dextrose tryptone agar (DTA): Use a commercial medium and follow the preparation and sterilization instructions provided by the manufacturer.
   (c) Sterile 2% agar (or seal agar): Weigh 20 g of agar and dissolve in 1 L of distilled water. Heat the mixture to boiling (*see* **Note 7**) to dissolve the agar. Dispense the

solution into media bottles and sterilize at 121 °C for 15 min.

2. Equipment:
   (a) Incubator capable of maintaining a uniform temperature of 50–55 °C.
   (b) Thermostatically controlled water or oil baths set at 55–60 °C (tempering agar), 100 °C, or 110 °C.
   (c) Colony counter.
3. Other tools:
   (a) Sterile bottle with a capacity of ≥500 mL, with a 100 mL volume marking.
   (b) Sterile Erlenmeyer flasks, 300 mL.
   (c) Sterile pipettes with capacities of 10 mL, along with a pipette bulb for liquid handling.
   (d) Sterile Petri plates (90 × 15 mm).
   (e) Thermometer.

### 2.4 Enumeration of Thermophilic Anaerobic Spores

1. Culture media and diluent:
   (a) Sterile distilled water.
   (b) Sterile sulfite agar: Weigh 10 g of tryptone, 1 g of sodium sulfite ($Na_2SO_3$), and 20 g of agar and dissolve in 1 L of water, mixing well. Place a clean iron strip or nail into each tube (20 × 150 mm) and dispense 10 mL in each tube. There is no need to adjust the pH of the medium. Autoclave at 121 °C for 15 min and then cool it to 55 °C before use.
   (c) Sterile and de-aerated PE-2 medium (*see* **Item 2.2(g)**).
   (d) Sterile and de-aerated liver broth (LB): Weigh 500 g of chopped beef liver, dissolve in 1 L of water, and bring to a slow boil (*see* **Note 7**) for 1 h. Adjust the pH to approximately 7.0, and then boil (*see* **Note 7**) for an additional 10 min. Strain the boiled mixture through cheesecloth, collect the liquid, and dilute it to a final volume of 1 L. To this broth, add 10.0 g of peptone and 1.0 g of $K_2HPO_4$, adjusting the pH again to 7.0 if necessary. Transfer 1–2 cm of preboiled ground beef liver into test tubes (20 × 150 mm) and add 10–12 mL of the prepared broth. Sterilize the tubes by autoclaving at 121 °C for 15 min. If the medium is not freshly prepared, heat it to boiling (*see* **Note 7**) for 20 min to remove oxygen and allow it to cool to 55 °C before use.
   (e) Sterile 2% agar (or seal agar, *see* **Item 2.3(c)**).
   (f) Sterile Vaspar (*see* **Item 2.2(j)**).

2. Equipment:
   (a) Incubator capable of maintaining a uniform temperature of 50–55 °C or 55 °C.
   (b) Thermostatically controlled water or oil baths set at 45 °C (tempering agar) and 100 °C.
   (c) Colony counter.
3. Other tools:
   (a) Sterile bottle with a capacity of ≥500 mL, with a 100 mL volume marking.
   (b) Sterile pipettes with capacities of 20 mL, along with a pipette bulb for liquid handling.
   (c) Sterile tubes (20 × 150 mm).
   (d) Water and ice bath.

## 3 Methods

### 3.1 Enumeration of Mesophilic Aerobic Spores

(a) Weigh 50 g of the sample into a sterile bag or container and then mix it with 450 mL of sterile 0.1% peptone water.

(b) Homogenize the mixture using a stomacher at normal or high speed for 30 to 60 s (for the bag) or manually agitate (for the container) to ensure proper dispersion of the sample.

(c) Transfer aliquots of 10 mL, 1 mL, and 0.1 mL from the blended mixture into three separate 500 mL Erlenmeyer flasks, each containing 100 mL of molten TGEA prewarmed and maintained at 45 °C.

(d) Gently swirl the flasks to ensure thorough dispersion of the sample in the medium.

(e) Place the flasks in a thermostatically controlled water or oil bath at 80 °C for 30 min. If using a flask with a capacity different from 500 mL, the 30-min thermal shock (*see* **Note 1**) should start when the medium reaches 80 °C, and temperature rise should be monitored with a thermometer using a control flask.

(f) After the thermal shock, rapidly cool the flasks in tepid water (~30–40 °C) to stop the thermal action, ensuring the temperature does not fall below the point where the agar begins to solidify.

(g) Pour the entire 100 mL contents from each flask into five sterile Petri plates (approximately 20 mL per plate).

(h) After the medium solidifies, invert the plates, and incubate them at 35 °C for 48 h under aerobic conditions.

(i) After incubation, count the colonies formed both on the surface and within the agar of the five Petri plates. Calculate the spore count per gram of the sample based on the dilution factors:

- Plates containing 10 mL of blended sample:

  Spores/*g* = Total number of colonies in the 5 Petri plates

- Plates containing 1.0 mL of blended sample:

Spores/*g* = 10 × Total number of colonies in the 5 Petri plates

- Plates containing 0.1 mL of blended sample:

Spores/*g* = 100 × Total number of colonies in the 5 Petri plates

**Rope-Producing Spores**

To enumerate only rope-producing spores, follow the method described above with the following modifications:

1. Use only the 10 mL and 1 mL aliquots of the initial mixture.
2. Use DTA culture medium in 250 mL Erlenmeyer flasks.
3. Subject the mixture to thermal shock by boiling (*see* **Note 7**) (94–100 °C) for 15 min, starting from the moment the temperature reaches 94 °C in a control flask.

A schematic representation of the procedure for the enumeration of mesophilic aerobic spores is shown in Fig. 1.

### *3.2 Enumeration of Mesophilic Anaerobic Spores*

(a) Weigh 25 g of the sample into a sterile bag or container; then, mix it with 225 mL of sterile distilled water or 0.1% peptone water.

(b) Homogenize the mixture using a stomacher at normal or high speed for 30 to 60 s (for the bag) or manually agitate (for the container) to ensure proper dispersion of the sample (*see* **Note 2**).

(c) Place the container in a thermostatically controlled water or oil bath under the following conditions, ensuring the temperature rise is monitored using a thermometer in a control flask. Start timing only once the target temperature is reached:

- More resistant spores (proteolytic): 80 °C for 10–13 min (*see* **Note 1**)
- More sensitive spores (nonproteolytic): 60 °C for 15–30 min (*see* **Note 1**)
- Spores with unknown resistance: 71 °C for 10 min (*see* **Note 1**).

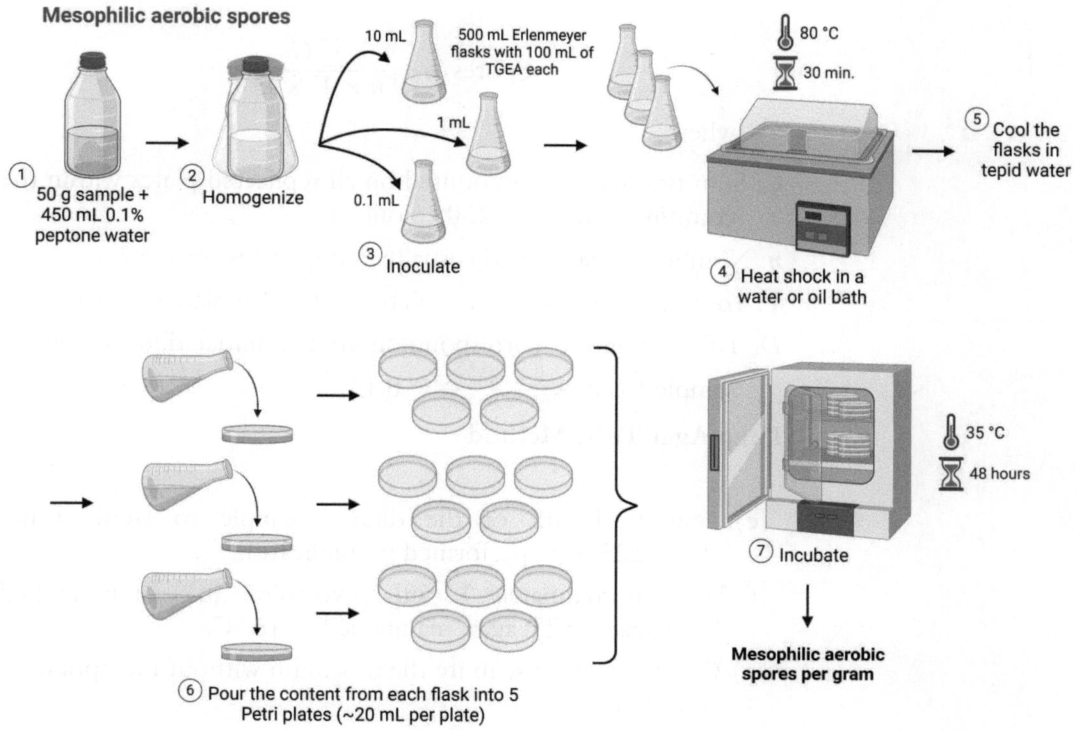


**Fig. 1** Schematic representation of the enumeration method for mesophilic aerobic spores. Created in BioRender [18]

(d) Rapidly cool the container in a water and ice bath to stop the thermal action.

**Plate Count Method**

(e) Transfer 1 mL of the diluted sample to sterile Petri plates (performed in duplicate).

(f) Pour approximately 20 mL per plate of premelted TPGY agar, TCD agar, AC agar, or TP agar, maintained at 45 °C.

(g) Gently swirl in a figure-eight motion to ensure uniform distribution of the sample.

(h) Allow the culture medium to solidify.

(i) Pour a thioglycollate agar overlay (*see* **Note 3**).

(j) Wait for the overlay to solidify.

(k) Incubate the plates inverted at 30–35 °C for 2–5 days under anaerobic conditions using anaerobic jars and anaerobic atmospheric generators.

(l) After incubation, count the colonies within the agar and calculate the results as "spores per gram of sample" using the equation:

$$\text{Spores}/g = \frac{\sum C}{n \times V \times D}$$

where:

*C*: Sum of the colonies counted on all replicated plates within the counting range (25–250 colonies).

*n*: Number of plates used (usually in duplicates, so $n = 2$).

*V*: Volume of the sample inoculated onto each plate (1 mL).

*D*: Dilution factor corresponding to the initial dilution of the sample $\left(D = \frac{25g}{(25g + 225\ mL)} = 0.1\right)$.

**Deep Agar Tube Method**

(e) Transfer 1 mL of the diluted sample to sterile tubes (16 × 125 mm, performed in duplicate).

(f) Pour approximately 12 mL (two-thirds full) of premelted TCD agar, or TP agar, maintained at 45 °C.

(g) Gently mix to distribute the inoculum without incorporating air bubbles.

(h) Cool and solidify the agar by immersing the tubes in cold water.

(i) Pour a 3–5 mL overlay of thioglycollate agar to create an oxygen barrier (*see* **Notes 3** and **6**).

(j) Wait for the overlay to solidify.

(k) Incubate the tubes at 30–35 °C for 2–3 days.

(l) Count the colonies directly in the tube and report the result as "spores per gram of sample" using the equation from Subheading 5.1.8.

**Most Probable Number (MPN) Method**

(e) For each dilution, transfer 1 mL or 0.1 mL to a series of three or five tubes (20 × 150 mm) containing PE-2 or CMM medium.

(f) Overlay the medium with petroleum jelly or Vaspar (*see* **Notes 3** and **6**).

(g) After the overlay solidifies, incubate the tubes at 30–35 °C for 2–3 days (for rapid germination and growth) and up to 1–3 weeks (for slow germination and growth).

(h) Determine the number of positive tubes per dilution (turbidity or gas formation) and express the result as the most probable number per gram (*MPN/g*) using the following equation:

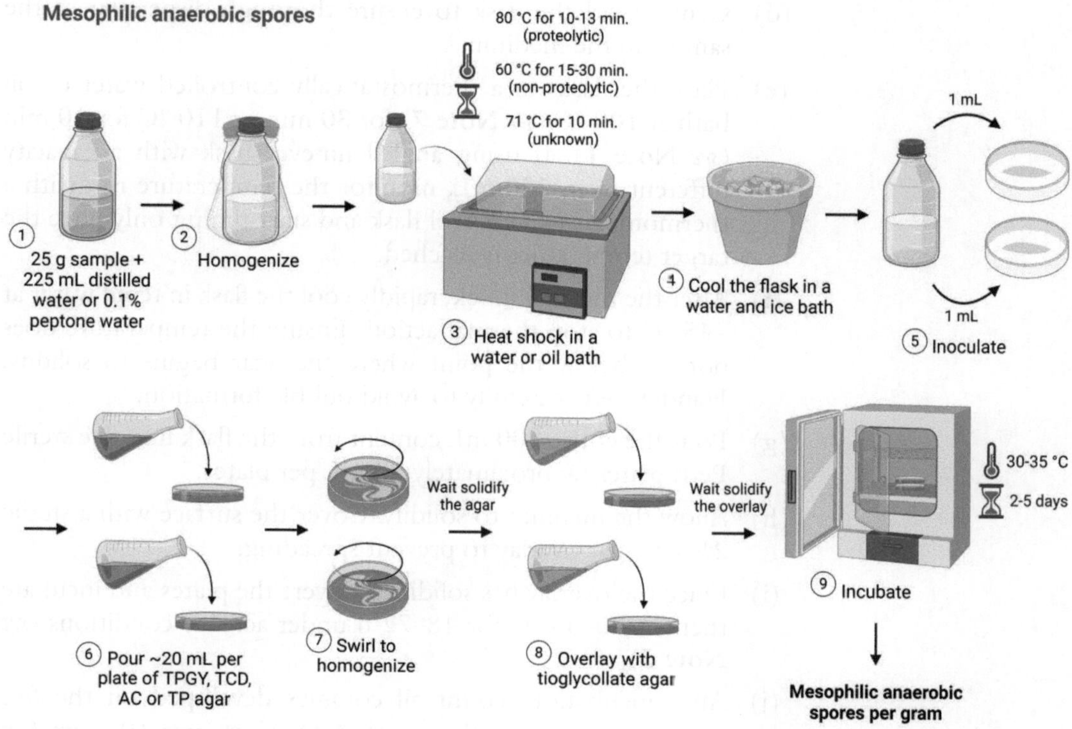


**Fig. 2** Schematic representation of the enumeration method for mesophilic anaerobic spores. Created in BioRender [18]

$$MPN/g = \frac{P}{\sqrt{NT}}$$

where:

$P$ = Number of positive tubes.

$N$ = Sum of the amount of sample inoculated (in grams) in all negative tubes.

$T$ = Sum of the amount of sample inoculated (in grams) in all tubes.

A schematic representation of the procedure for the enumeration of mesophilic anaerobic spores is shown in Fig. 2.

**3.3 Enumeration of Thermophilic Aerobic Spores**

(a) Weigh 20 g of the sample into a sterile flask with a mark of 100 mL volume. Pour sterile distilled water to reach the 100 mL mark.

(b) Homogenize the mixture by agitation to ensure proper dispersion of the sample (*see* **Note 4**).

(c) Transfer a 10 mL aliquot from the blended mixture into a 300 mL Erlenmeyer flask containing 100 mL of sterile molten DTA maintained at 55–60 °C.

(d) Gently swirl the flask to ensure thorough dispersion of the sample in the medium.

(e) Place the flasks in a thermostatically controlled water or oil bath at 100 °C (*see* **Note 7**) for 30 min or 110 °C for 10 min (*see* **Note 1**). If using an Erlenmeyer flask with a capacity different from 300 mL, monitor the temperature rise with a thermometer in a control flask and start timing only once the target temperature is reached.

(f) After the thermal shock, rapidly cool the flask in tepid water at ~45 °C to stop thermal action. Ensure the temperature does not fall below the point where the agar begins to solidify. Handle flasks carefully to avoid bubble formation.

(g) Pour the entire 100 mL content from the flask into five sterile Petri plates (approximately 20 mL per plate).

(h) Allow the medium to solidify. Cover the surface with a sterile 2% seal agar overlay to prevent spreading.

(i) Once the overlay has solidified, invert the plates and incubate them at 50–55 °C for 48–72 h under aerobic conditions (*see* **Note 5**).

(j) After incubation, count all colonies developed on the five plates and calculate the number of spores per 10 g or 1 g using the following equations:

$$\text{Spores}/10g = \frac{\sum C}{D}$$

Or,

$$\text{Spores}/g = \frac{\sum C}{10 \times D}$$

*C*: Total number of colonies in the five Petri plates.

*D*: Dilution factor, corresponding to the initial dilution applied to the sample $\left( D = \frac{20g}{100\ mL} = 0.2 \right)$.

**Flat Sour Spores**

If the goal is to enumerate only flat sour thermophilic spores (capable of acidifying the medium without gas production):

1. Count only colonies with the following characteristics: round shape, 2 to 5 mm in diameter; dark, opaque center and yellow halo surrounding the colony against a purple medium background (typical of flat sour spores).
2. Calculate the number of flat sour spores per 10 g or 1 g using the same equations provided above.

A schematic representation of the procedure for the enumeration of thermophilic aerobic spores is shown in Fig. 3.

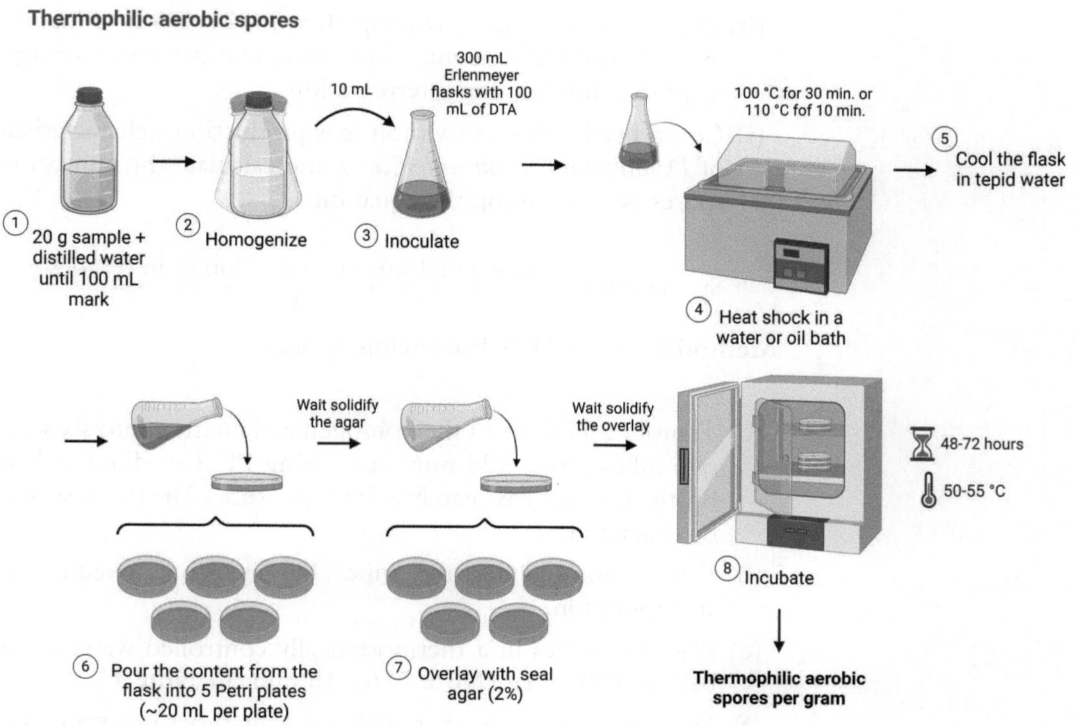


**Fig. 3** Schematic representation of the enumeration method for thermophilic aerobic spores. Created in BioRender [18]

### *3.4 Enumeration of Thermophilic Anaerobic Spores*

(a) Weigh 20 g of the sample into a sterile flask marked with 100 mL volume. Pour sterile distilled water to reach the 100 mL mark.

(b) Homogenize the mixture manually to ensure proper dispersion of the sample (*see* **Note 4**).

**Method for $H_2S$-Producing Spores**

(c) Transfer 20 mL from the homogenized mixture into six separate tubes (20 × 150 mm) containing sulfite agar (approximately 3.3 mL per tube). The agar should be molten, deaerated, and maintained at 45 °C. Note: Deposit the inoculum below the surface of the medium.

(d) Gently homogenize the tubes by inversion, avoiding air incorporation.

(e) Place the tubes in a thermostatically controlled water or oil bath at 100 °C (*see* **Note 7**) for 15 min (*see* **Note 1**).

(f) After the thermal shock, rapidly cool the tubes in a water and ice bath to stop thermal action.

(g) Allow the agar to solidify and incubate the tubes at 50–55 °C for 24–48 h.

(h) Perform a preliminary reading after 20–24 ± 3 h of incubation, as some strains may cause complete darkening or agar rupture, complicating interpretation.

(i) Count black colonies (without gas production), characteristic of *Desulfotomaculum nigrificans*, and calculate the number of spores per 10 g using the equation:

$$Spores/10g = \frac{10 \times \text{Total number of colonies in 6 tubes}}{4}$$

**Method for Non-$H_2S$-Producing Spores**

(c) Transfer 20 mL from the homogenized mixture into six separate tubes (16 × 125 mm) containing PE-2 medium or liver broth (LB), approximately 3.3 mL per tube. The broth should be deaerated.

(d) Gently homogenize the tubes by inversion, avoiding air incorporation.

(e) Place the tubes in a thermostatically controlled water or oil bath at 100 °C (*see* **Note 7**) for 15 min (*see* **Note 1**).

(f) After the thermal shock, rapidly cool the tubes in a water and ice bath to stop thermal action.

(g) Cover the surface of the medium with 3–5 mL of seal agar (2%) or warm Vaspar (~45 °C) to create anaerobic conditions (*see* **Notes 3** and **6**).

(h) Allow the seal agar or Vaspar to solidify and then incubate the tubes at 55 ± 2 °C for 72 h.

(i) After incubation, evaluate positive and negative tubes. Tubes with positive growth will exhibit signs such as gas production, upward displacement of the seal agar, or color change (from purple to yellow).

(j) Using the number of positive tubes, express the results as the Most Probable Number per gram (*MPN/g*) using the following equation:

$$MPN/g = \frac{P}{\sqrt{NT}}$$

where:

$P$ = Number of positive tubes.

$N$ = Sum of the amount of sample inoculated (in *g*) in all negative tubes.

$T$ = Sum of the amount of sample inoculated (in *g*) in all tubes.

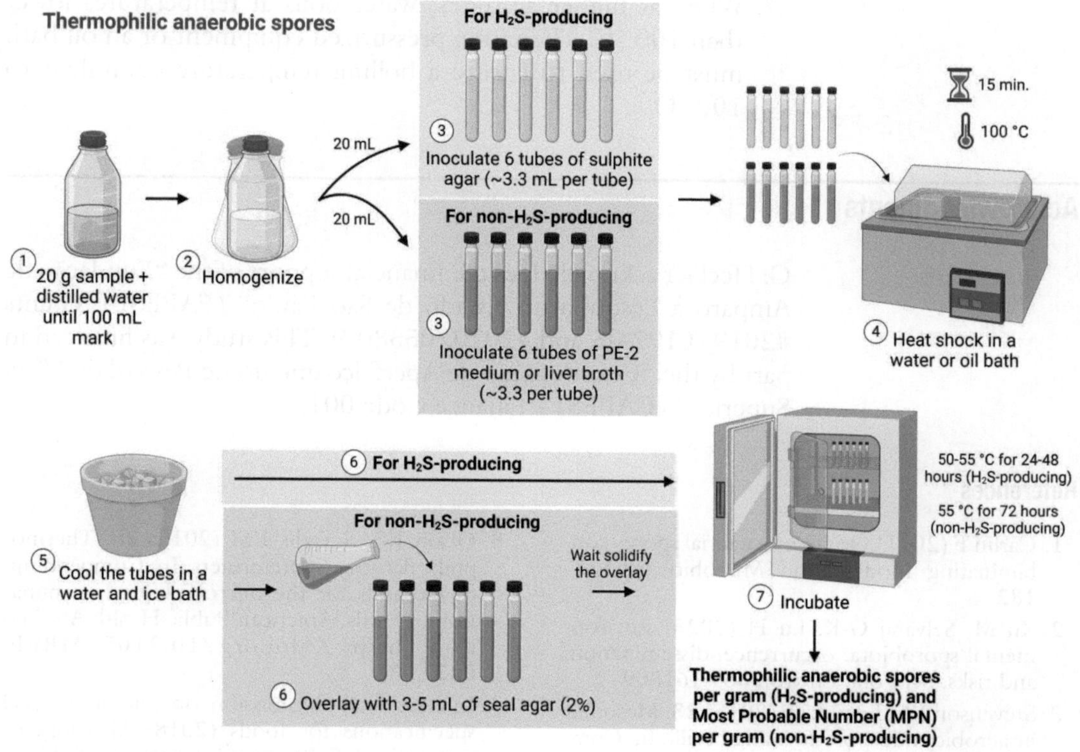


**Fig. 4** Schematic representation of the enumeration method for thermophilic anaerobic spores. Created in BioRender [18]

A schematic representation of the procedure for the enumeration of thermophilic anaerobic spores is shown in Fig. 4.

## 4 Notes

1. This step eliminates vegetative cells and activates spores.
2. If necessary, perform serial dilutions in sterile tubes or bottles with the diluent.
3. The use of overlays reduces oxidation and promotes spore growth under anaerobic conditions.
4. Glass beads can be added to assist in sample homogenization.
5. To prevent dehydration of agar plates when incubating at 55 °C, maintaining a humidity level of 50% to 70% in the incubator is recommended.
6. This step can be performed in an ice water bath to facilitate the solidification of the seal agar.

7. When at higher altitudes, water boils at temperatures lower than 100 °C. Therefore, pressurized equipment or an oil bath must be used to ensure a boiling temperature equivalent to 100 °C.

## Acknowledgments

C. Heckler acknowledges the financial support of the "Fundação de Amparo à Pesquisa do Estado de São Paulo" (FAPESP): Grants #2019/01974-5 and #2020/05580-9. This study was financed in part by the "Coordenação de Aperfeiçoamento de Pessoal de Nível Superior" (CAPES) – Finance Code 001.

### References

1. Carlin F (2011) Origin of bacterial spores contaminating foods. Food Microbiol 28:177–182
2. Xu M, Selvaraj G-K, Lu H (2023) Environmental sporobiota: occurrence, dissemination, and risks. Sci Total Environ 869:161809
3. Stevenson KE, Lembke F (2015) 23. Mesophilic aerobic endospore-forming bacilli. In: Compendium of methods for the microbiological examination of foods. American Public Health Association. https://doi.org/10.2105/MBEF.0222.028
4. Vos P, Garrity GM, Jones D, Krieg NR, Ludwig W, Rainey FA, Schleifer K-H, Whitman WB (eds) (2009) Bergey's manual of systematic bacteriology: the firmicutes, vol 3. Springer, New York, NY
5. Pereira APM et al (2020) Occurrence and enumeration of rope-producing spore forming bacteria in flour and their spoilage potential in different bread formulations. LWT 133:110108
6. Pacher N et al (2022) Ropiness in bread—a re-emerging spoilage phenomenon. Foods 11:3021
7. Tortorelli S, Anderson JE (2015) 24. Mesophilic anaerobic Sporeformers. In: Compendium of methods for the microbiological examination of foods. American Public Health Association. https://doi.org/10.2105/MBEF.0222.029
8. Olson KE, Sorrells KM (2015) 26. Thermophilic flat sour Sporeformers. In: Compendium of methods for the microbiological examination of foods. American Public Health Association. https://doi.org/10.2105/MBEF.0222.031
9. International commission on microbiological specifications for foods (2018) Microorganisms in foods 7. Springer International Publishing, Cham. https://doi.org/10.1007/978-3-319-68460-4
10. Enache E, Podolak R (2015) 27. Thermophilic anaerobic Sporeformers. In: Compendium of methods for the microbiological examination of foods. American Public Health Association. https://doi.org/10.2105/MBEF.0222.032
11. Ronner AB, Elliott PH (2015) 28. Sulfide Spoilage Sporeformers. In: Compendium of methods for the microbiological examination of foods. American Public Health Association. https://doi.org/10.2105/MBEF.0222.033
12. Salfinger Y, Tortorello ML (eds) (2015) Compendium of methods for the microbiological examination of foods. American Public Health Association. https://doi.org/10.2105/MBEF.0222
13. Blodgett R (2023) BAM Appendix 2: most probable number from serial dilutions. In: Bacteriological Analytical Manual (BAM)
14. Andrews WH, Hammack TS (2023) BAM Chapter 1: food sampling/preparation of sam-

ple homogenate. In: *Bacteriological Analytical Manual (BAM)*

15. AOAC Official Method 977.26 (2023) *Clostridium botulinum* and its toxins in foods. In: Official methods of analysis of AOAC International. Oxford University Press, New York. https://doi.org/10.1093/9780197610145.003.2235
16. AOAC Official Method 972.45 (2023) Thermophilic bacterial spores in sugars. In: Official methods of analysis of AOAC International. Oxford University Press, New York. https://doi.org/10.1093/9780197610145.003.2233
17. AOAC Official Method 972.44 (2023) Microbiological method. In: Official methods of analysis of AOAC International. Oxford University Press, New York. https://doi.org/10.1093/9780197610145.003.2231
18. Heckler C (2024) Spore's illustration. Created in BioRender. https://BioRender.com/f00h551

# Chapter 2

# Enumeration of Heat-Resistant Microorganisms: Fungi

**Dionisio Pedro Amorim Neto and Beatriz Dal Pian Machado**

**Abstract**

Fungal spores, particularly heat-resistant molds, are a significant challenge throughout the food production chain, from raw materials to processed products. These resilient microorganisms can lead to food spoilage, local immunological reactions, and serious health risks, such as mycotoxin poisoning. Therefore, effective prevention and control strategies, coupled with regular monitoring, are crucial to avoid and contain heat-resistant mold contamination. In this sense, this chapter outlines a protocol using the pour plate technique to quantify these molds, offering detailed steps and adaptations for broader applications.

**Key words** Heat-resistant molds, Thermoresistant, Ascospores, Quantification

## 1 Introduction

The ubiquitous presence of fungal spores [1], particularly in environments such as soil where fruits and vegetables are cultivated [2], makes molds a constant challenge in food preservation. Among fungal spores, ascospores (sexual reproduction cycle) stand out for their unusual heat resistance, far exceeding that of conidia (asexual reproduction cycle) [3, 4]. The resistance of ascospores to temperatures near 80 °C allows them to survive pasteurization, high-pressure processing, and other thermal treatments, ensuring their dissemination [5, 6]. In the case of heat-resistant molds, heat, instead of inhibiting, stimulates the germination process of their spores [3, 7], making them even more resilient and challenging to control throughout the food production chain. This characteristic of ascospores gives molds the ability to contaminate the entire food production chain, from cultivation, through surfaces and utensils, to processed products such as juices, purees, baby food, and soft drinks [3, 5]. The fungal genera *Byssochlamys*, *Thermoascus*, *Neosartorya*, *Talaromyces*, and *Eupenicillium* are known for their ability to withstand high temperatures [1], making them significant targets in the food industry. Beyond food spoilage, these heat-resistant

Anderson S. Sant'Ana (ed.), *Heat Resistant Microorganisms in Food Safety and Spoilage*, Methods and Protocols in Food Science, https://doi.org/10.1007/978-1-0716-5280-0_2,

fungi pose a wide range of health risks. They can trigger allergic reactions and pulmonary infections and, in some cases, produce mycotoxins such as verruculogen, patulin, and ochratoxin A, which pose a serious threat to food safety [6]. The implementation of effective prevention and control strategies, coupled with continuous monitoring of fungal contamination levels, is essential to mitigate the impact of these resilient microorganisms on food safety and the economic stability of the production chain. Aiming to contribute to this mitigation, this chapter details a protocol for the quantification of heat-resistant molds using the pour plate technique. This method, consolidated by the Compendium [8], is the result of theoretical and practical development from other classical works [9–12] on the enumeration of heat-resistant fungi. In addition to presenting a step-by-step protocol (Fig. 1), this chapter also discusses variations and adaptations found in the literature, aiming to broaden the scope and possibilities of applying this methodology.

## 2 Materials

### 2.1 Media

1. *Malt Extract Agar (MEA)* in g/L: 20 g of malt extract, 20 g of dextrose, 1.0 g of bacteriological peptone, 20 g of bacteriological agar, and 1 L of distilled water. Under stirring, adjust the final pH to 5.4 ± 0.2 using 1 N HCl. Sterilize by autoclaving at 15 lb pressure (121 °C) for 15 min. To prepare a double-strength medium, multiply the quantity of each ingredient by 2 while keeping the total volume of water at 1 L. Once solidified, the culture medium can be stored under refrigeration (4 °C) for a maximum period of 30 days. On the day of analysis, it should be melted in a microwave. Alternatively, immediately after autoclaving, the medium can be kept in a liquid state in an incubator at 50–55 °C but for not more than 24 h.

### 2.2 Reagents and Solutions

1. *Distilled water*: Sterilize by autoclaving at 15 lb pressure (121 °C) for 15 min. Can be stored at room temperature.

## 3 Methods

### 3.1 Sampling and Dilution

1. Weigh 100 g of sample and mix it with 150 mL of sterile distilled water (*see* **Note 1**) in a stomacher bag. For juices and other specific samples, *see* **Notes 2–5**.
2. Blend the sample and water thoroughly in the stomacher for up to 4 min.

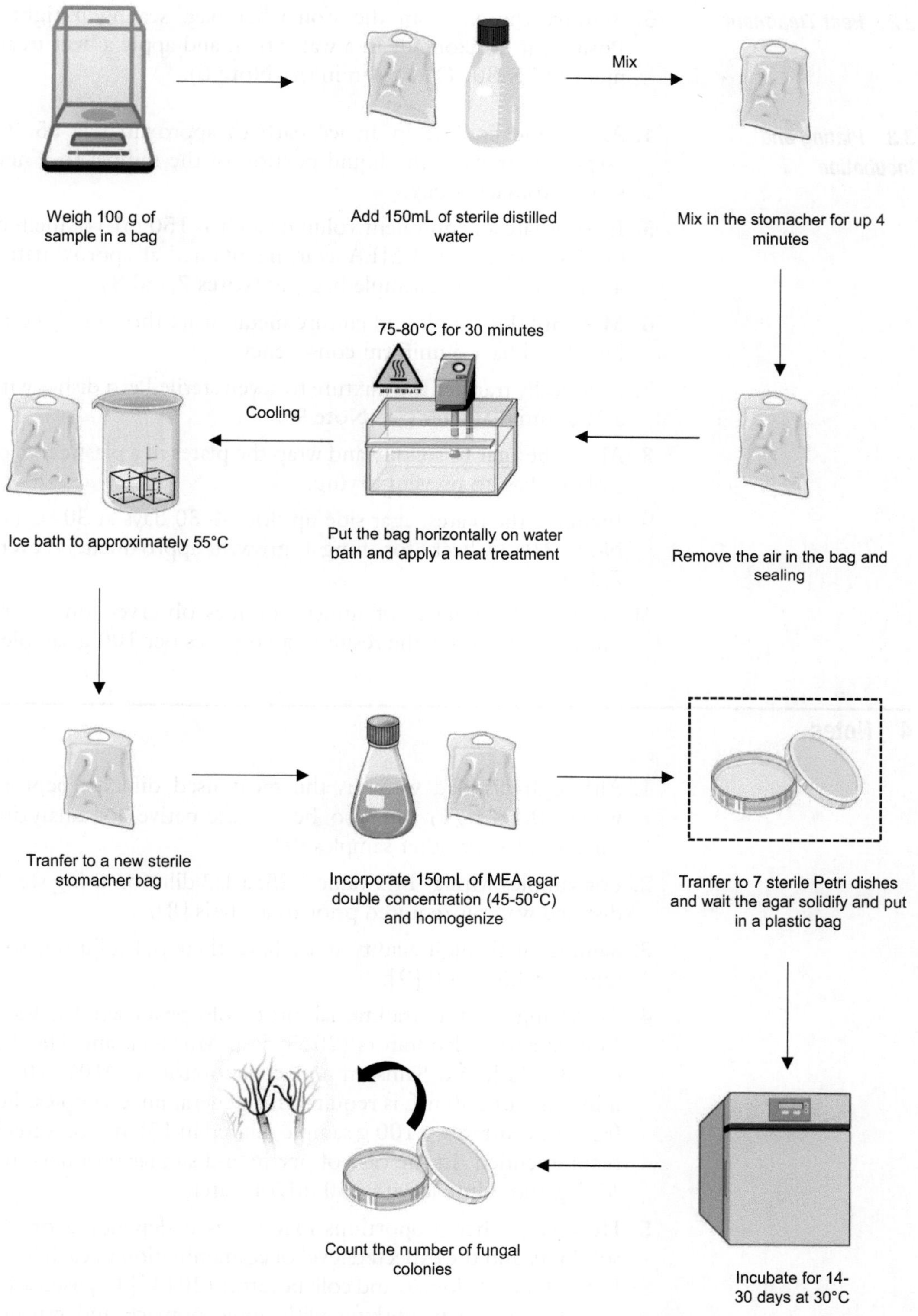


**Fig. 1** Experimental procedure for counting spores of heat-resistant fungi in food (This figure was created using icons and vector images obtained and adapted from "© 2016 DBCLS TogoTV, CC-BY-4.0 https:/creativecommons.org/licenses/by/4.0/, licensed under a Creative Commons Attribution 4.0 Unported License)

### 3.2 Heat Treatment

3. Remove the air from the stomacher bag, sealing it tightly. Position it horizontally in a water bath and apply a heat treatment of 75–80 °C for 30 min (*see* **Note 6**).

### 3.3 Plating and Incubation

4. After rapid cooling in an ice bath to approximately 55 °C, aseptically transfer the liquid portion of the sample to a new sterile stomacher bag.
5. Incorporate an equivalent volume (such as 150 mL) of melted, double-concentrated MEA agar, maintained at approximately 45–50 °C, into the sample bag (*see* **Notes 7** and **8**).
6. Mix until the sample and culture medium are thoroughly combined and have a uniform consistency.
7. Aseptically transfer the mixture to seven sterile Petri dishes with a 150 mm diameter (see **Note 9**).
8. Allow the agar to solidify and wrap the plates in a plastic film or a plastic bag to prevent drying.
9. Incubate the plates, agar side up, for 14–30 days at 30 °C (*see* **Note 10**). Check for fungal growth approximately every 7 days.
10. Count the number of fungal colonies observed on all the plates and express the result as ascospores per 100 g sample.

## 4 Notes

1. Although distilled water is the most used diluent, peptone water (0.1% w/v) can also be an alternative for analyzing surfaces [13] or other samples [9].
2. For samples with a °Brix value ≥35, a 1:2 dilution using sterile distilled water is required prior to analysis [8].
3. Samples with high acidity must have their pH adjusted to a range of 3.5 to 4.0 [9].
4. According to the traditional protocols proposed by Rico-Munoz and collaborators (2015) [8], Murdock and Hatcher (1978) [12], and Samsom and collaborators (2010) [10], a minimum of 250 mL is required for general juice samples. For fruit concentrates, a 100 g sample diluted in 150 mL of water is recommended. In the case of pectin and similar products, use 12.5 g and dilute it with 250 mL of water.
5. However, other proportions may be used depending on the solubility and the expected level of contamination in each food. For instance, Salomão and collaborators (2014) [14] used a 1:1 (w/v) ratio when working with apple pomace and crushed apple. Tranquillini and colleagues (2017) [15], on the other hand, adopted a 100 g sample size for water-soluble extracts

and lemon cells and a 500 g sample size for berries. Blueberry samples were prepared at a 1:1 (w/w) ratio using 200 g of sample [16], while mixed fruit jams were diluted at a ratio of 10 g to 15 mL of water [17]. Due to the soil's high capacity to retain spores, a larger dilution of 1:24 (w/v) was necessary for soil samples [18].

6. The studies examined in this chapter employed a heat treatment protocol of 80 °C for 30 min when analyzing foods [14–17, 19] and surfaces [13]. However, for soil samples, a less stringent condition of 75 °C for 20 min was utilized [18].
7. A variety of approaches have been used with Malt Extract Agar [20, 21], the most common culture medium in this protocol [8, 9]. While some studies have used the medium without supplementation [15], others have added chloramphenicol (200 mg/L) [17, 19] or a combination of chloramphenicol (100 mg/L) and rose bengal (50 mg/L) [18] to inhibit bacterial and excessive fungal growth.
8. Another culture medium recommended for this assay [8, 9] is the *Potato Dextrose Agar (PDA)* in g/L: 4 g of potato infusion (infusion from 200 g potatoes), 20 g of dextrose, 15 g of bacteriological agar [8, 22], pH 5.5. Previous studies have used PDA in different formulations: Double-strength, acidified with 10% (v/v) tartaric acid [13], supplemented with rose bengal and chloramphenicol [14], or commercially as *PGRB*, containing only rose bengal [16].
9. The number of pour plates used can range from 4 to 8 [9, 14, 15, 17], with each plate having a diameter of around 150 mm.
10. Typically, incubation is carried out at 30 °C for either 14 [8, 18] or 30 days [9, 13–15, 17, 19]. Nevertheless, the incubation temperature and duration can be modified to suit specific experimental requirements. For instance, incubating at 25 °C for 10 days [16] is a suitable option under certain circumstances.

## References


1. Pahalagedara ASNW, Gkogka E, Hammershøj M (2024) A review on spore-forming bacteria and moulds implicated in the quality and safety of thermally processed acid foods: focusing on their heat resistance. Food Control 166: 110716. https://doi.org/10.1016/J.FOODCONT.2024.110716
2. Rico-Munoz E (2017) Heat resistant molds in foods and beverages: recent advances on assessment and prevention. Curr Opin Food Sci 17: 75–83. https://doi.org/10.1016/J.COFS.2017.10.011
3. Rico-Munoz E, Samson RA, Houbraken J (2019) Mould spoilage of foods and beverages: using the right methodology. Food Microbiol 81:51–62. https://doi.org/10.1016/J.FM.2018.03.016
4. Snyder AB, Martin N, Wiedmann M (2024) Microbial food spoilage: impact, causative agents and control strategies. Nature reviews microbiology 22(9):528–542. https://doi.org/10.1038/s41579-024-01037-x
5. Ishara AWS, Gunasena GDDK (2021) Heat resistant Moulds in pasteurized fruit syrups.

Eur J Agric Food Sci 3:104–111. https://doi.org/10.24018/EJFOOD.2021.3.1.199

6. Frac M, Jezierska-Tys S, Yaguchi T (2015) Occurrence, detection, and molecular and metabolic characterization of heat-resistant fungi in soils and plants and their risk to human health. Adv Agron 132:161–204. https://doi.org/10.1016/BS.AGRON.2015.02.003
7. Pinto CA, Moreira SA, Fidalgo LG et al (2020) Effects of high-pressure processing on fungi spores: factors affecting spore germination and inactivation and impact on ultrastructure. Compr Rev Food Sci Food Saf 19:553–573. https://doi.org/10.1111/1541-4337.12534
8. Rico-Munoz E, Houbraken J, Samson RA (2015) Detection and enumeration of heat-resistant molds. In: Compendium of methods for the microbiological examination of foods. American Public Health Association
9. Pitt JI, Hocking AD (2009) Fungi and food spoilage. In: Fungi and Food Spoilage, pp 1–519. https://doi.org/10.1007/978-0-387-92207-2/COVER
10. Samson RA, Houbraken J, Thrane U et al (2010) Food and indoor fungi: Second Edition. CBS-KNAW Fungal Biodiversity Centre
11. Houbraken J, Samson RA (2006) Standardization of methods for detecting heat resistant fungi. Adv Exp Med Biol 571:107–111. https://doi.org/10.1007/0-387-28391-9_5
12. Murdock DI, Hatcher WS (1978) A simple method to screen fruit juices and concentrates for heat-resistant Mold. J Food Prot 41:254–256. https://doi.org/10.4315/0362-028X-41.4.254
13. Delgado DA, de Sant'Ana AS, de Massaguer PR (2012) Occurrence of molds on laminated paperboard for aseptic packaging, selection of the most hydrogen peroxide- and heat-resistant isolates and determination of their thermal death kinetics in sterile distilled water. World J Microbiol Biotechnol 28:2609–2614. https://doi.org/10.1007/S11274-012-1064-8/FIGURES/1
14. Salomão B d CM, Muller C, do Amparo HC, de Aragão GMF (2014) Survey of molds, yeast and Alicyclobacillus spp. from a concentrated apple juice productive process. Braz J Microbiol 45:49–58. https://doi.org/10.1590/S1517-83822014000100008
15. Tranquillini R, Scaramuzza N, Berni E (2017) Occurrence and ecological distribution of Heat Resistant Moulds Spores (HRMS) in raw materials used by food industry and thermal characterization of two Talaromyces isolates. Int J Food Microbiol 242:116–123. https://doi.org/10.1016/J.IJFOODMICRO.2016.11.023
16. Fan L, Martynenko A, Doucette C et al (2018) Microbial quality and shelf life of blueberry purée developed using cavitation technology. J Food Sci 83:732–739. https://doi.org/10.1111/1750-3841.14073
17. Santos JLP dos, Samapundo S, Biyikli A et al (2018) Occurrence, distribution and contamination levels of heat-resistant moulds throughout the processing of pasteurized high-acid fruit products. Int J Food Microbiol 281:72–81. https://doi.org/10.1016/J.IJFOODMICRO.2018.05.019
18. Maneeboon T, Sangchote S, Hongprayoon R et al (2023) Occurrence of heat-resistant Mold Ascospores in pineapple and sugarcane field soils in Thailand. Int J Microbiol 2023:8347560. https://doi.org/10.1155/2023/8347560
19. Alam M, Sarkar A, Sayem ASM et al (2024) Accessing the growth of heat-resistant mold ascospores in potato dextrose agar, isolated from a processed fruit jam—dependence of temperature, pH, and sugar concentration. Food Sci Nutr 12(11):9371–9378. https://doi.org/10.1002/FSN3.4486
20. Reddish G (1919) A substitute for beer wort as a bacteriological culture medium. J Bacteriol 3:06–07
21. Rapp M (1974) Indikatorzusätze zur Keimdifferenzierung auf Würze- und Malzextrakt-Agar. Milchwiss 29:341–344
22. Beever RE, Bollard EG (1970) The nature of the stimulation of fungal growth by potato extract. Microbiol (N Y) 60:273–279. https://doi.org/10.1099/00221287-60-2-273

# Chapter 3

# Determining the Cause of Spoilage of Heat-Processed Foods

**Larissa Pereira Margalho**


## Abstract

Determining the cause of spoilage in heat-processed foods is critical for ensuring food safety, quality, and compliance with regulatory standards. This step-by-step protocol provides a comprehensive guide for identifying the root cause of spoilage, encompassing both microbial and nonmicrobial factors. The protocol begins with collecting and documenting spoiled samples, followed by analytical steps, including visual inspection, sensory evaluation, and microbiological testing to detect potential pathogens or spoilage organisms. Chemical and physical analyses assess pH, hydrogen gas, and packaging integrity. The protocol also emphasizes reviewing processing parameters, such as time–temperature profiles, and evaluating storage conditions. By systematically isolating and identifying the contributing factors, this protocol enables food manufacturers to implement targeted corrective actions, prevent future occurrences, and maintain product quality. This method applies to various heat-processed food products, including canned goods, pasteurized items, and retort-processed foods.


**Key words** Canned food, Container integrity, Microbial troubleshooting, Commercial sterility test, Post-process contamination, Spore-forming bacteria

## 1 Introduction

Heat processing is one of the most widely used methods for preserving food, ensuring safety, and extending shelf life. Techniques such as pasteurization, sterilization, and canning effectively destroy or inhibit the growth of microorganisms, making foods safe for consumption and preventing spoilage [1, 2]. However, spoilage can still occur in heat-processed foods, posing challenges to food safety and quality.

Spoilage in heat-processed foods is typically caused by several factors, including inadequate processing, post-process contamination, or the survival of heat-resistant microorganisms such as thermophilic bacteria, bacterial spores, and fungi [3, 4]. These microorganisms can lead to physical, chemical, or sensory changes in the food, affecting its color, texture, odor, and overall appearance, ultimately rendering it unfit for consumption [5]. Some

Anderson S. Sant'Ana (ed.), *Heat Resistant Microorganisms in Food Safety and Spoilage*, Methods and Protocols in Food Science, https://doi.org/10.1007/978-1-0716-5280-0_3,

bacterial spores, such as *Clostridium botulinum* and *Bacillus cereus*, can survive thermal treatments and germinate under favorable conditions, leading to spoilage and potential foodborne illnesses [6].

Understanding the causes and mechanisms of spoilage in heat-processed foods is essential for improving food preservation techniques and ensuring product safety. Research has focused on detecting spoilage microorganisms, determining the conditions that promote their growth, and developing strategies to mitigate spoilage through optimized heat treatments and improved packaging systems [7, 8]. By addressing these challenges, the food industry can enhance product quality and safety, meeting consumer expectations and regulatory standards.

This protocol aims to systematically determine the cause of spoilage in heat-processed foods by identifying potential microbial, chemical, and physical factors. The findings from such investigations can help refine processing conditions and develop more effective food preservation strategies.

## 2 Materials

All necessary materials, media, and reagents are summarized in Table 1 and divided per section.

## 3 Methods

### 3.1 Selection and Visual Inspection of Container Samples

1. Use a representative number of suspecting spoilage samples ($n = 4$ to 6) and at least two for controls for the same batch.
2. Record product details: Note the product type, size, brand, batch number, expiration date, and processing details (e.g., retort temperature, time, and pressure).
3. Carefully remove the label from the container and file it as a document. Mark the container with a permanent marker with the sample process code so that it can be accurately tracked.
4. Examine the exterior of the container for physical defects, such as dents, rust, swollen cans, vacuum loss, leaking seams, broken seals, perforations, and imperfections, to check for any apparent defects. If an ink-jet code is applied, check it for any color variation indicating the retort was made.
5. Note any irregularities, such as alterations in the container's color, consistency, or texture. Check the labels for leakage stains and highlight them if found. Retain the labels as records for future assessment. Dispose of containers that show clear signs of damage.

**Table 1**
**Materials, equipment, and medium required to perform tests to determine the cause of spoilage of heat-processed foods**

| Section | Material and equipment | Medium | Reagents |
|---|---|---|---|
| 3.1 Selection and visual inspection of container samples | Permanent marker, notebook for recording product details, sterile gloves, magnifying glass or microscope (for detailed inspection), light source (for checking imperfections) | – | Adhesive remover |
| 3.2 Weighing and preparing containers | Analytical balance, sterile paper towels, flame-sterilized utensils (scalpel, scissors, can opener, knife), sterile forceps/tweezers, sterile tray | – | Sporicidal disinfectant |
| 3.3 Qualitative hydrogen analysis | Sterile tray, sterile paper towels, sterile syringe and needle, sterile test tubes, Bunsen burner (for flame test), sterilized ethanol wipes | – | – |
| 3.4 Opening containers aseptically | Laminar flow hood, Bunsen burner, sterile materials (trays, containers, sanitary can opener, forceps/tongs, wipes, tongs, scalpel or knife, scissors), gloves, safety goggles | – | Sterile alcohol (70%)<br>Sterile disinfectant solution |
| 3.5 Aseptic sample collection | Laminar flow hood, sterile plastic bags, sterile sample containers, refrigeration unit, sterile collection utensils (forceps, spoons, spatula, knives), sterile materials (pipette, syringe, strainer, large-bore pipette, test tubes), gloves, and safety goggles | – | – |
| 3.6 Odor and appearance records | Odor reference guide, color and texture reference chart, notebook for recording observations, gloves (to prevent cross-contamination) | – | – |
| 3.7 Measuring pH | pH meter, disposable pipettes, sterile flask for pH measurement of the sample | Sterile saline solution | pH calibration buffers |

(continued)

**Table 1**
**(continued)**

| Section | Material and equipment | Medium | Reagents |
| --- | --- | --- | --- |
| 3.8 Microscopy examination | Microscope slides, cover slips, phase-contrast/fluorescence microscope, sterile inoculation loop | – | Gram staining reagents, malachite green staining, immersion oil |
| 3.9 Culturing-based microbiology analysis | | | |
| 3.9.1 Sample preparation and maintenance | Incubators (30 °C, 35 °C, 55 °C), autoclave, laminar flow hood, sterile test utensils (tubes, flasks, petri dishes, graduated pipettes, tips), sterile plastic bags, sterile paper towels, sterile wipes | Peptone water (0.1%) (in case of 1:1 dilution) | – |
| 3.9.2 Low-acid food (pH >4.5) | Sterile marking pen, sterile forceps/tweezers, Bunsen burner, anaerobic jar, incubators (30 °C, 35 °C, 55 °C), sterile workbench, sterile test tubes, sterile petri dishes, sterile Durham tubes, boiling water bath (100 °C), ice bath, | Cooked Meat Medium (CMM), PE-2 broth, Dextrose Tryptone Broth (DTB), Tryptic Soy Agar (TSA) plates, Liver Veal Agar (LVA) plates, Manganese Nutrient Agar (MnNA) plates, agar solution (1.5–2%), Vasparagar (1:1 jelly petroleum:xxx) | – |
| 3.9.3 Acid or acidified foods (pH <4.5) | Incubators (25 °C, 30 °C, 35 °C, 43 °C, 55 °C), boiling water bath (100 °C), ice bath, autoclave, anaerobic jar, sterile utensils (test tubes, petri dishes, graduated pipettes, flasks, inoculating loops, forceps/tweezers, syringes and needles, spoons/spatulas, and plastic bags) | Thermoacidurans Broth (TAB), APT broth, MRS broth, OS broth, ALI broth, acidified PDA plates, TAA plates, DTA plates, K agar plates, agar solution (1.5–2%), Vasparagar (for anoxic environment), sterile saline solution (0.85%), microaerophilic environment generation kit | – |

### 3.2 Weighing and Preparing Containers

1. Certify that the container is free of labels, adhesive, and dried food residue. Then, weigh the whole content to the nearest gram or ounce. Record the weight data.
2. Clean and sanitize the container using a sporicidal disinfectant. Soak the area to be opened with the solution (following the manufacturer's recommendations) for 10–15 min. This can be done by (I) placing the area to be opened on a tray containing the disinfectant solution or (II) using sterile paper towels soaked in the solution. Alcohol use is not recommended as it may not sterilize the entire surface completely. Thus, dry the

area with a sterile towel paper and open it using an autoclaved-sterilized utensil (e.g., sharp blades, scissors, scalpels, and bacteriological can opener, depending on the type of container). Use forceps or tweezers to manipulate the container and keep it aseptic.

### 3.3 Qualitative Hydrogen Analysis

1. This step is applied only to the swollen containers.
2. The container should be placed in a sterile tray and covered with sterile paper towels or ethanol wipes to avoid eruption or overflow while opening to prevent cross-contamination.
3. Carefully puncture the container and collect the gas using an inverted test tube perpendicularly placed onto the puncture point using the sterile needle.
4. After collection, immediately place the opening of the tube over a flame. Use a flame test in a controlled environment. Hydrogen gas is flammable and will produce a "pop" sound when ignited, confirming its presence. Record this information.

   4.1 Alternatively, a sterilized syringe and needle can be used (either by sterilizing them in an autoclave or by rinsing them with 70% ethanol and allowing them to air-dry). After sterilization, a sterile needle can be used to carefully puncture the lid of the container. The syringe can then be inserted into the can's headspace, and 5–10 mL of gas can be slowly withdrawn. As previously described, the collected gas can be transferred into a sterile tube and burned in a flame. This method is more appropriate for flexible or semi-rigid containers.
5. In the case of positive results, the sample must be quickly opened for further culturing tests.

### 3.4 Opening Containers Aseptically

#### 3.4.1 Metal Cans

1. Place the can on a stable, flat surface within the laminar flow hood or near the Bunsen burner. To avoid surface contamination, keep it inside a sterile tray.
2. Use the sterile sanitary can opener to puncture the lid. Open the can away from your body to avoid injury. Leave 0.64–1.27 cm of the end in contact with the double seam to keep it intact for subsequent quality control tests, such as micro-leak and container examination.
3. Slowly lift the lid using sterile tongs or forceps to release the built-up pressure.

#### 3.4.2 Glass Jars

1. Use sterile tongs to grip the lid and twist it counterclockwise. If the lid is difficult to open, tap the edges gently with a sterile spoon to break the vacuum seal.

*3.4.3 Plastic or Semi-Rigid Containers*

1. Use sterile scissors or a knife to carefully cut open the package, avoiding contact with the contents. Choose an area away from the sealing area. It can be from the bottom or side to prevent damage.
2. Make a hole large enough to collect the sample aseptically using a suitable, sterilized, sharp tool, such as a knife tip, scalpel, or scissors.

### 3.5 Aseptic Sample Collection

1. Collect at least two samples (~50 g) in sterile plastic bags or suitable containers—one designated for subculturing and the other retained as a reference control. Label each with the corresponding sample code for proper identification.
2. Transfer approximately 10 g of each sample to the sterile bag/container. Keep the reference sample under refrigeration. Depending on the physical state of the food samples, follow the following steps:

   2.1. *Solid Foods (e.g., Canned Meats, Vegetables)*

   1. Use sterile forceps, spoons, spatula, or knives to collect a representative portion from different areas of the food (top, middle, and bottom) and transfer it to a sterile container/bag.
   2. Avoid touching the food with your hands or nonsterile objects.
   3. Collect portions from different layers to ensure sample representativity for layered products (e.g., canned lasagna, pâté).

   2.2. *Liquid Foods (e.g., Soups, Broths, Juices)*

   1. Shake the container gently to ensure homogeneity.
   2. Use a sterile pipette or syringe to withdraw the required volume, transfer it into a sterile container, and seal it to prevent contamination.

   2.3. *Particulate Foods (Mixture of Small-Size Solids and Liquids)*

   1. If separation is needed, use a sterile strainer to isolate solid particles.
   2. Collect samples of both the solid and liquid phases separately in sterile containers.
   3. If the food has multiple components (e.g., chunks in sauce), sample the solid and liquid parts separately. You can do this using a large-bore pipette or by breaking the tips' ends.
   4. If analyzing them together, mix them carefully in a sterile container before sealing.

2.4. *Semi-Solid or Paste-Like Foods (e.g., Baby Food, Purées, Condensed Milk)*

1. Use a sterile spatula to transfer the sample into a sterile container.
2. If the product has variations in texture, collect from multiple areas to ensure an accurate representation.

### 3.6 Odor and Appearance Records

1. Record the characteristics of odor and appearance of the product, taking into consideration the following descriptive terms:

1.1. Odor: Acid, butyric acid, cheesy, fecal, fermented, medicinal, metallic, musty, pungent, putrid, rotten egg, sharp, sour, sweet

1.2. Texture and consistency: Tough, rigid, soft, slimy, digested, fluid, viscous, ropy, liquefied, coagulated

1.3. Color: Lighter or darker than the control

1.4. Appearance: Cooked vs. uncooked; turbidity or precipitated

1.5. Gas formation: Frothy or slight bubbles

### 3.7 Measuring pH

1. Separate a secondary sample to measure pH in an aseptic container. Make sure the secondary sample is not contaminated.
2. Measure the sample's pH using a pH meter or indicator strips. Ensure the equipment is calibrated, and the protocol for the food being analyzed is correct and standard.
3. Compare the pH with the expected range for the food product. An abnormal pH may indicate microbial spoilage. Report all data.

### 3.8 Microscopy Examination

1. Prepare a bacterial smear from the food product.
2. Perform Gram staining to differentiate bacteria.
3. Use phase contrast or fluorescence microscopy to detect spores and fungi.
4. Observe the smear under a microscope and note the morphology (e.g., rods, cocci, etc.), motility, and presence of endospores.
5. Perform malachite green staining to observe bacteria endospores.

1. All precautions reported before must be addressed before microbiological tests begin.
2. Considering the shelf life of the product to be analyzed:

### 3.9 Culturing-Based Microbiology Analysis

#### 3.9.1 Sample Preparation and Maintenance

(a) Samples with less than 30 days of shelf life from the fabrication date: pre-incubate at 30/35 °C for 10 days before analysis. If destined for storage at >40 °C, the samples should also be incubated at 55 °C for 7 days.

(b) Samples with more than 30 days of shelf life from fabrication date: Do not need pre-incubation.

(c) Altered samples: Samples showing swelling, leakage, or any apparent defects do not need pre-incubation. Keep it in a resistant, sterile plastic bag and under refrigeration until analysis. If the container is opened, transfer the content to a sterile flask.

3. Open the container aseptically and collect the samples, as recommended in Subheadings 3.2, 3.4, and 3.5.

#### 3.9.2 Low-Acid Food (pH >4,5)

1. Inoculate 1 mL or 1 g of food sample in four tubes containing 9 mL of CMM or PE-2 broth and four tubes of DTB. Ensure the CMM or PE-2 tubes were previously de-aerated (heated at 100 °C for 20 min and cooled to 45 °C, without agitation). A Duhan tube should be placed in all tubes for gas formation assessment.
2. Cover the CMM and PE-2 tubes with 2–5 mL of agar solution (1.5–2%) or Vasparagar (to create an anoxic environment).
3. Streak a loop of food onto two TSA plates for control.
4. Incubate the tubes and plates under the following conditions:
   - CMM/PE-2 tubes with seal agar → 30 or 35 °C for up to 7 days under normal atmosphere. Target microorganism: Anaerobic mesophilic spore-forming bacteria.
   - CMM/PE-2 tubes with seal agar → 55 °C for up to 7 days under normal atmosphere. Target microorganism: Anaerobic thermophilic spore-forming bacteria.
   - DTB tubes → 30 or 35 °C for up to 7 days under normal atmosphere. Target microorganism: Aerobic mesophilic spore-forming bacteria.
   - DTB tubes → 55 °C for up to 7 days under normal atmosphere. Target microorganism: Aerobic thermophilic spore-forming bacteria.
   - TSA plate → 30 or 35 °C for up to 7 days under normal atmosphere. Control for aerobic mesophilic spore-forming bacteria growth. The aim is to ensure that there is no cross-contamination in the laboratory.
   - TSA plate → 30 or 35 °C for up to 7 days under anaerobiosis. Control for anaerobic mesophilic spore-forming bacteria growth.

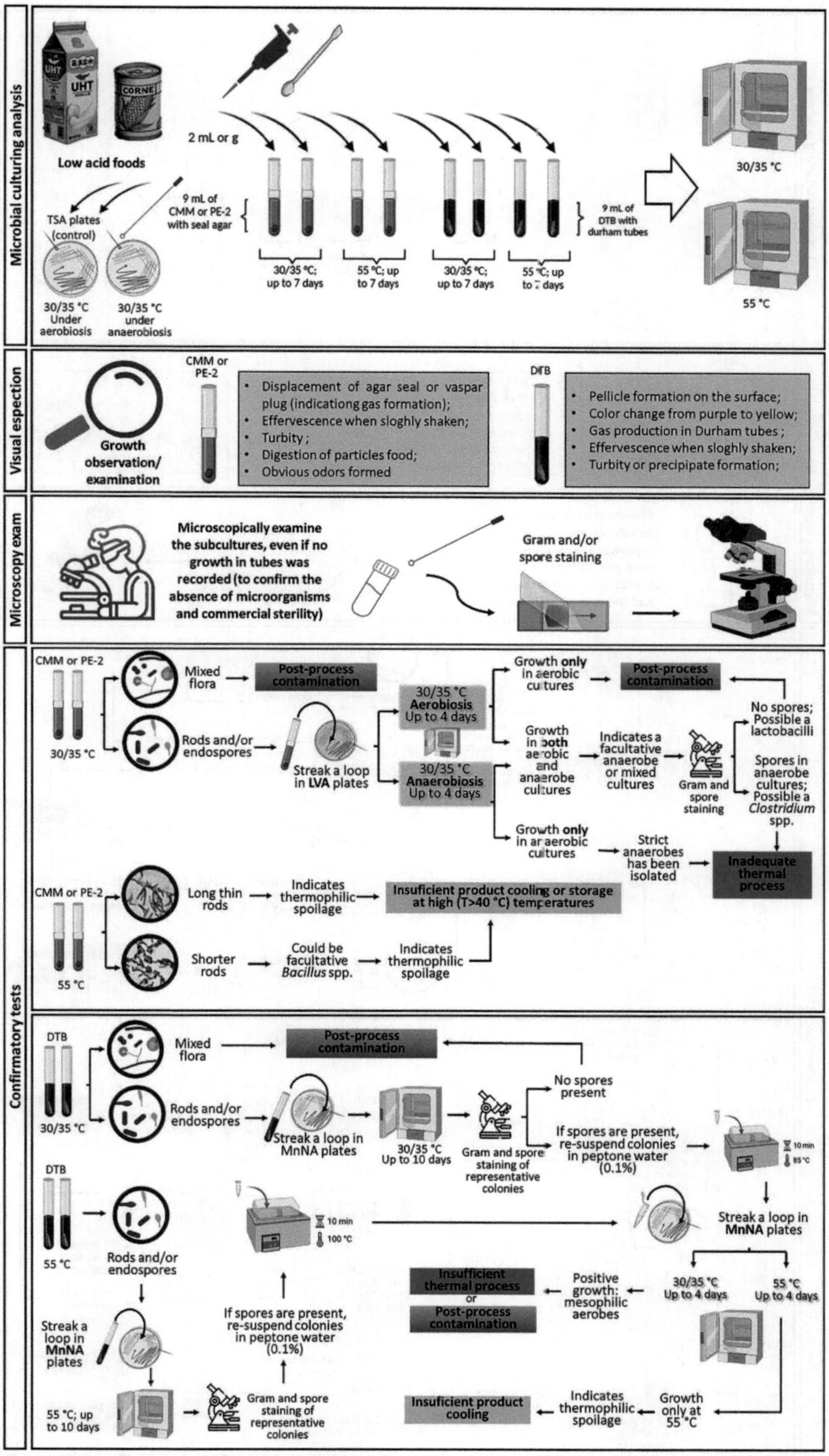


**Fig. 1** Flowchart for culturing-based microbiology analysis of low acid foods (pH >4.5), confirmatory tests for aerobic and anaerobic cultures and possible cause of spoilage data interpretation (Fig. 2)

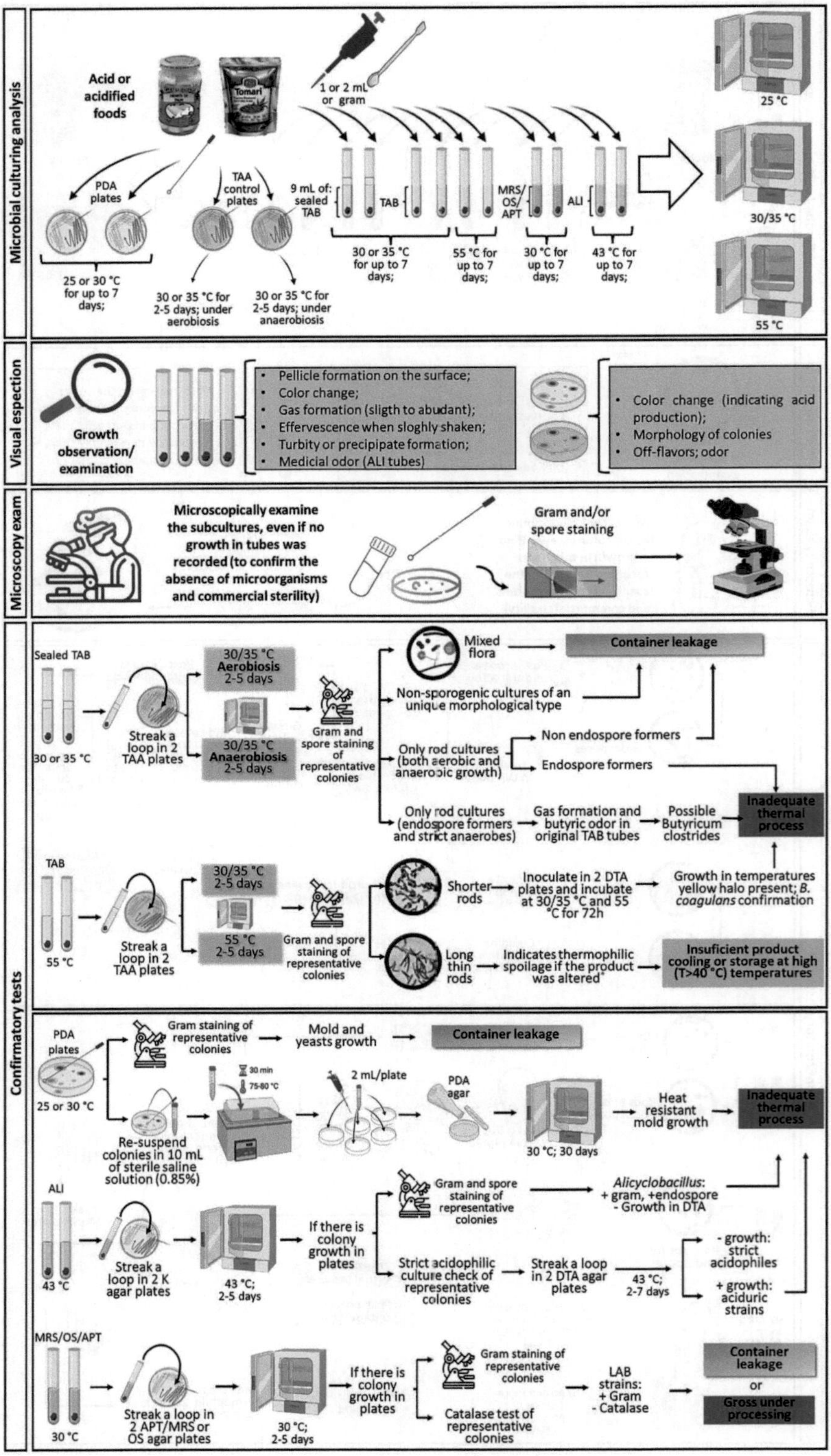


**Fig. 2** Flowchart for culturing-based microbiology analysis of acid or acidified foods (pH <4.5), confirmatory tests for aerobic and anaerobic cultures and possible cause of spoilage data interpretation

5. In case of apparent growth in tubes (indicated by turbidity, color change, or gas formation), proceed with microscopy examination and the confirmatory tests as follows:
   - CMM/PE-2 tubes with seal agar → Inoculate a loop from each tube in two Liver Veal Agar (LVA) plates by streaking it on the surface. Incubate one plate in aerobic conditions and the other in anaerobic conditions for 4 days at the same temperature of the original tube.
     - In the case of colony growth, a microscopic assay will be performed to check the morphology of strains and/or the presence of endospores.
   - DTB tubes → Inoculate a loop from each tube in one Manganese Nutrient Agar (MnNA) plate by streaking it on the surface. Incubate one plate in aerobic conditions and the other in anaerobic conditions for up to 10 days at the same temperature of the original tube. After incubation, select different colony types to perform morphological tests (spore and gram staining). Regarding spore formation in the plates incubated at 55 °C, the following tests will be performed to classify the colonies as strict or facultative thermophiles.
     - *Strict or facultative thermophiles*: Using a sterile inoculating loop, transfer a small amount of bacterial culture (obtained in the 55 ° C MnNA plates) into 1 mL of sterile peptone water tubes (0.1%). Submit the tubes to thermal shock for 10 min at boiling temperature and cool them immediately in an ice bath. Streak the content of tubes in two new plates of MnNA and incubate one plate at 35 °C/4 days and the other at 55 °C/4 days. Growth at both temperatures indicates the presence of facultative thermophiles. Growth at 55 °C only shows the presence of strict thermophiles.
     - *Strict mesophiles or facultative thermophiles*: Using a sterile inoculating loop, transfer a small amount of bacterial culture (obtained in the 30/35 ° C MnNA plates) into 1 mL of sterile peptone water tubes (0.1%). Submit the tubes to thermal shock for 10 min at 85 °C and cool them immediately in an ice bath. Streak the content of tubes in two new plates of MnNA and incubate one plate at 35 °C/4 days and the other at 55 °C/4 days. Growth at both temperatures indicates the presence of facultative thermophiles. Growth at 35 °C only shows the presence of strict mesophiles.
6. Interpret the results following the step-by-step flowchart presented in Fig. 1. Note all visual and microexam observations.

*3.9.3 Acid or Acidified Foods (pH <4.5)*

1. Inoculate 1–2 mL or g of the food sample into six tubes containing 9 mL of *Thermoacidurans* Broth (TAB), two tubes with APT, MRS, or OS broth, and two tubes with ALI broth.
2. Cover the two tubes of TAB with 2–5 mL of Agar seal solution (1.5–2%) or Vaspar (to create an anoxic environment).
3. Streak a loopful of food onto two acidified PDA and two TAA plates for control.
4. Incubate the tubes and plates in the following conditions:
   - 2 TAB + seal agar/Vaspar tubes → 30 or 35 °C for up to 7 days under normal atmosphere. Target microorganism: Anaerobic mesophilic aciduric bacteria.
   - 2 TAB tubes without agar seal → 30 or 35 °C for up to 7 days under normal atmosphere. Target microorganism: Aerobic and facultative anaerobic mesophilic aciduric bacteria.
   - 2 TAB tubes without agar seal → 55 °C for up to 7 days under normal atmosphere. Target microorganism: Aerobic thermophilic aciduric bacteria.
   - 2 APT/MRS or OS broth tubes → 30 °C for up to 7 days under normal atmosphere. Target microorganism: Lactic acid bacteria.
   - 2 ALI broth tubes → 43 °C for up to 7 days under normal atmosphere. Target microorganism: *Alicyclobacillus* spp.
   - 2 Acidified PDA plates → 25 or 30 °C for up to 7 days under normal atmosphere. Target microorganisms: Molds and yeasts.
   - 1 TAA plate → 30 or 35 °C for up to 5 days under normal atmosphere. Control for aerobic mesophilic aciduric bacteria growth.
   - 1 TAA plate → 30 or 35 °C for up to 5 days under anaerobic atmosphere. Control for anaerobic mesophilic aciduric bacteria growth.
5. In the case of apparent growth in tubes (indicated by turbidity, color change, or gas formation), proceed with microscopy examination and the confirmatory tests as follows:
   - 2 TAB + seal agar/vaspar tubes → Inoculate a loopful from the tubes onto two TAA plates by streaking. Incubate one plate under aerobic conditions and the other under anaerobic conditions, maintaining the same temperature as in the original tube. If bacterial growth is observed, perform morphological characterization using Gram staining on representative colonies and examine under a microscope. If endospores are present, conduct spore staining for further identification.

- 2 TAB (55 °C) → Inoculate a loopful from the tubes onto two TAA plates by streaking. Incubate one plate at 30/35 °C and the other at 55 °C for 2–5 days. If bacterial growth is observed, perform morphological characterization as previously described.
  - *Bacillus coagulans confirmation*: The rod-shaped colonies should be inoculated in two plates of DTA. Incubate on plate at 30/35 °C and the other at 55 °C for 72 h.
- 2 Acidified PDA plates → When evidence of mold and yeast growth is observed, perform a test for:
  - *Morphological characterization*: If presumptive yeast and mold growth are observed, morphological characterization will be performed as previously described.
  - *Thermoresistant filamentous fungi*: Suspend a loopful of each mold in 10 mL of sterile saline solution (0.85%) tubes. Heat the tubes at 75–80 °C for 30 min. Distribute the 10 mL into five sterile petri dish plates (2 mL per plate) and pour acidified PDA medium. Incubate the plates at 30 °C for 30 days. To avoid medium dehydration, cover the plates in sterile plastic bags.
- 2 APT/MRS or OS broth tubes → Inoculate a loopful from the tubes onto two APT/MRS or OS agar plates by streaking. Incubate the plates at 30 °C for 2–5 days under a microaerophilic atmosphere. If colony growth is observed, perform the following tests for at least five representative colonies:
  - Gram staining
  - Catalase test
- 2 ALI broth tubes → Inoculate a loopful from the tubes onto two K agar plates by streaking. Incubate the plates at 43 °C for 2–5 days. If colony growth is observed, perform the Gram and endospore staining for at least five representative colonies.
  - *Confirmation of strict acidophilic strains*: Inoculate a loopful from the tubes onto two DTA agar plates by streaking. Incubate the plates at 43 °C for 2–7 days.

## 4 Disposal of Contents

1. If food samples present a risk of contamination by pathogens and/or toxins, transfer them to another container and sterilize them by autoclaving.

2. If samples do not show growth of microorganisms, dispose of them appropriately as organic waste.

## 5 Cleaning and Examining Containers

1. If samples present a risk of contamination by pathogens or toxins, wash the container carefully using appropriate handling equipment. Then, collect all the washing water in a suitable container and sterilize it by autoclaving.
2. Remove all food residue from the container, wash with soap and water (if necessary), and disinfect with a sporicidal solution. Allow the container to dry at 35 °C for later detailed exam analysis.
3. If using a brush, be careful to keep the container's integrity.
4. Visually analyze the container's interior for unusual aspects, such as discoloration, seal imperfections, holes, cracks, cuts, etc. Consult a specialist for a deeper evaluation.

## 6 Further Key Results Interpretation

For a more detailed investigation, consult Supplementary material (Tables 1 and 2), adapted from Tables 62.4 and 62.5 of the Compendium [9].

## 7 Notes

1. Collect as much information as possible (when available) about the samples concerning the thermal process applied (temperature x time), storage conditions, and number of defective packages per total number of samples processed. For transparent containers, examine and note any abnormal appearance (turbidity, gas formation, under-filling, digestion, etc.). All of this contributes to a better conclusion of the analysis.
2. Avoid examining extremely swollen containers or dented cans. If this is not an option, refrigerate them for a few hours before opening to prevent eruption. However, this procedure can destroy some vegetative thermophilic strains.
3. To avoid cross-contamination, the environment, materials, utensils, and analysts must be kept clean from the moment of reception throughout the analysis. The workbenches must be sanitized with a sporicidal disinfectant. These precautions will ensure the reliability of the tests and prevent false-positive results.

**Table 2**
**Didactic guide to identifying causes of spoilage in acid or acidified canned foods (pH $\leq$ 4.6)**

| Diagnosis | Container condition | Odor | Appearance of contents | Gas detected | pH | Microbiological examination |
|---|---|---|---|---|---|---|
| Hydrogen swells | Swollen | Normal to metallic | Normal to frothy (cans usually etched or corroded) | >20% $H_2$ | Normal | Negative to few microorganisms |
| Post-process contamination or grossly insufficient thermal process | Swollen | Sour | Frothy; possibly ropy brine | Mostly $CO_2$ | Below normal | Rods, coccoids, cocci, or yeasts (pure or mixed cultures) |
| No process given | Swollen | Sour | Frothy; possibly ropy brine; food particles firm | Mostly $CO_2$ | Below normal | Rods, coccoids, cocci, or yeasts (pure or mixed cultures); if high exhaust, only sporeformers recovered |
| Insufficient cooling (thermophilic spoilage) | Swollen (in foods with normal pH 4.0–4.6) | Normal to sour-cheesy | Frothy | $H_2$ and $CO_2$ | Normal | Rods, medium short to medium long, usually granular; spores seldom seen |
| Insufficient processing (butyric acid anaerobes) | Swollen (in foods with normal pH 4.0–4.6) | Butyric acid | Frothy; large volume of gas | $H_2$ and $CO_2$ | Below normal | Rods with bipolar staining; possibly spores |
| Grossly insufficient processing (*Lactobacilli*) or post-process contamination | Swollen | Sour | Frothy | Mostly $CO_2$ | Below normal | Short to long rods |
| Insufficient vacuum | No vacuum and/or buckled | Normal | Normal | No $H_2$ | Normal to slightly below normal | Negative to moderate number of microorganisms |
| Insufficient processing (*B. coagulans*)—limited to tomato juice | Flat | Sour to "medicinal" | Normal | No $H_2$ | Normal to below normal | Rods, possibly granular in appearance |

(continued)

**Table 2**
**(continued)**

| Diagnosis | Container condition | Odor | Appearance of contents | Gas detected | pH | Microbiological examination |
|---|---|---|---|---|---|---|
| Insufficient cooling (thermophilic spoilage) | Flat | Medicinal or phenolic | Normal | No $H_2$ | Normal | Rods |
| Post-process contamination or no process given | Flat | Normal to sour | Normal to cloudy brine; possibly moldy | No $H_2$ | Slightly to below normal | Rods, coccoids, cocci, or mold (pure or mixed cultures) |

4. When opening the container, avoid cutting it in regions close to the double seam and side seams of cans, the sealing of lids in glass jars, and the heat seals of flexible and semi-rigid containers (such as plastic cans or jars, trays, pouches, bottles, etc.).
5. All utensils used to open the containers must be autoclaved since alcohol flaming is ineffective for sterilization.
6. In the microscopy examination, a control sample should be used to differentiate microorganisms from food particles.
7. If piping the samples directly from the container for the culturing-based microbiology analysis is impossible, dilute them in sterile water in a 1:1 ratio.
8. Only qualified individuals should perform microbiological analyses due to the risk of *Clostridium botulinum* growth in tubes incubated at 30–35 °C.
9. All broth and agar media used in Subheading 3.9.3 should be acidified to pH ≤4.6 or adjusted to closely match the product's pH, as only microorganisms capable of growing under these conditions are relevant for analysis.
10. If the growth of cultures in the tubes is uncertain, proceed with the standard analysis and microscopic examinations as usual. If no growth is observed in the subcultures, it can be concluded that no growth occurred in the original tube.

## Acknowledgments

L. P. Margalho acknowledges the financial support of the "Fundação de Amparo à Pesquisa do Estado de São Paulo" (FAPESP): Grant #2020/14631-6. This study was financed in part by the "Coordenação de Aperfeiçoamento de Pessoal de Nível Superior" (CAPES) – Finance Code 001.

### References

1. Jay JM, Loessner MJ, Golden DA (2005) Modern food microbiology. Springer
2. Doyle MP, Beuchat LR (2013) Food microbiology: fundamentals and frontiers. ASM Press
3. Peck MW (2006) Clostridium botulinum and the safety of minimally processed foods. Trends Food Sci Technol 17(1):48–55
4. André S, Vallaeys T, Planchon S (2017) Spore-forming bacteria responsible for food spoilage. Res Microbiol 168(4):379–387
5. Lund BM, Baird-Parker AC, Gould GW (2000) The microbiological safety and quality of food. Springer
6. Montville TJ, Matthews KR (2013) Food microbiology: an introduction. ASM Press
7. ICMSF (International Commission on Microbiological Specifications for Foods) (2005) Microorganisms in foods 6: microbial ecology of food commodities. Springer
8. Smid EJ, Lacroix C (2013) Microbial cultures for food fermentation. Annu Rev Food Sci Technol 4:123–140
9. Salfinger Y, Tortorello ML (eds) (2015) Compendium of methods for the microbiological examination of foods. American Public Health Association

# Chapter 4

# Conducting Test of Commercial Sterility


Leonardo do Prado-Silva and Caroline Heckler



## Abstract

Commercial sterility testing is applied to food products that undergo thermal treatment and are sealed in hermetic packaging (e.g., cans, aseptic packages, pouches, and glass jars) to verify the effectiveness of the thermal process. For this test, foods are classified based on their acidity: low-acid foods (pH > 4.6) and acid or acidified foods (pH ≤ 4.6). The objective of this chapter is to systematically describe the steps of the commercial sterility test for food products, ensuring the validation of product safety, reduction of noncompliance risks, and maintenance of consumer trust. This chapter serves as a foundational document for quality assurance teams, promoting consistency in testing practices and alignment with evolving regulatory standards.



**Key words** Low-acid foods, Acidified food, Botulism, Canned food, Hermetic seal


## 1 Introduction

Commercial sterility can be achieved through thermal processing of foods, combined with or without pH reduction and/or water activity ($a_w$) adjustment, which must be sufficient to ensure no microbial proliferation in the product [1]. Unlike absolute sterility, which implies complete elimination of microorganisms, commercial sterility refers to the absence of microorganisms capable of proliferating under defined storage and usage conditions [1]. Products deemed commercially sterile must be packaged in hermetic containers that provide a physical barrier against environmental microbial contamination [2, 3].

Potential microbial contaminants that may develop depend on the product's pH and water activity, as these parameters influence microbial growth. Consequently, products are classified into low-acid foods (pH > 4.6 and $a_w$ > 0.85) and acid or acidified foods (pH ≤ 4.6) [4]. Among the most significant low-acid foods are meat, fish, and dairy products, canned vegetables, and ready-to-eat soups. In these products, pH > 4.6 and $a_w$ > 0.85 are not restrictive, allowing the growth of most microorganisms, with


Anderson S. Sant'Ana (ed.), *Heat Resistant Microorganisms in Food Safety and Spoilage*, Methods and Protocols in Food Science, https://doi.org/10.1007/978-1-0716-5280-0_4,

*Clostridium botulinum* being the primary target of thermal processing due to its ability to cause botulism [5]. Conversely, prominent acidic foods include fruit juices, tomato derivatives, and acidified vegetables. Here, pH $\leq$ 4.6 acts as a growth-limiting factor for most microorganisms, except molds, yeasts, and aciduric bacteria (e.g., lactic acid bacteria, *Bacillus spp.*, *Clostridium spp.*, and *Alicyclobacillus spp.*) [6].

Commercial sterility testing enables food processing industries to mitigate risks and losses associated with microbial contamination. Contaminated products pose significant hazards, including foodborne illnesses and economic losses due to recalls or reputational damage. For example, in low-acid canned foods, surviving *C. botulinum* spores may lead to lethal botulism, while in acidic products, microbial contamination can result in economic losses due to spoilage from microbial activity [7]. Thus, regulatory agencies require the development of rigorous protocols to verify whether sterilization or aseptic processing methods achieve the intended level of microbial control.

This protocol outlines a systematic approach for commercial sterility testing in low-acid and acidic foods, designed to detect viable microorganisms in finished products using culture-based microbiological techniques. The primary objectives are to:

1. Ensure the efficacy of thermal processing by reducing microbial contamination to acceptable levels for each product type.
2. Verify compliance with regulatory requirements.
3. Identify microbial contamination, including spore-forming and non-spore-forming bacteria, molds, and yeasts.

## 2 Materials

### 2.1 Media

#### 2.1.1 Low-Acid Foods (pH > 4.6)

1. *Dextrose tryptone broth (DTB)**, in g/L: 10.0 g of tryptone, 5.0 g of dextrose, 0.04 g of bromocresol purple, and 1 L of distilled water. Dissolve the ingredients in distilled water by heating, if necessary, and dispense 20 mL portions into 20 × 200 mm screw-capped culture tubes. Sterilize by autoclaving at 121 °C for 15 min. Store at 4 °C for up to 30 days [8].
2. *Cooked meat medium (CMM)**, in g/L: 454.0 g of heart muscle, 10.0 g of peptone, 10.0 g of "Lab-Lemco" powder, 5.0 g of sodium chloride, 2.0 g of dextrose, and 1 L of distilled water. Dissolve the ingredients in distilled water by heating, if necessary, and dispense 20 mL portions into 20 mm × 200 mm screw-capped culture tubes. Sterilize by autoclaving at 121 °C for 15 min. Store at 4 °C for up to 30 days [8].

3. *PE-2 medium*, in g/L: 20.0 g of peptone, 3.0 g of yeast extract, 0.04 g of bromocresol purple, and 1 L of distilled water. Dissolve the ingredients in distilled water by heating, if necessary, and dispense 20 mL portions into 20 × 200 mm screw-capped culture tubes. Add 8 to 10 untreated pea seeds. After hydrating the pea seeds for 1 h, sterilize by autoclaving at 121 °C for 15 min [8].
4. *Nutrient agar* with manganese sulfate (NAMn). Preparation A*, in g/L: 15.0 g of agar, 5.0 g of peptone, 5.0 g of sodium chloride, 1.5 g of beef extract, 1.5 g of yeast extract, and 1 L of distilled water. *Preparation B*, in g/L: Dissolve 3.08 g of manganese sulfate ($MnSO_4$) in 100.0 mL of distilled water. Add 1.0 mL of Preparation B to Preparation A (nutrient agar) and sterilize by autoclaving at 121 °C for 15 min. Wait for the melted media to reach an approximate temperature of 55 °C before pouring it into sterile Petri plates (90 × 15 mm diameter). Store at 4 °C for up to 30 days [8].
5. *Liver veal agar (LVA)**, in g/L: 20.0 g of meat liver base, 11.0 g of agar, 1.2 g of sodium sulfite, 0.75 g of dextrose, 0.75 g of soluble starch, and 0.5 g of ammonium iron (III) citrate, and 1 L of distilled water. Sterilize by autoclaving at 121 °C for 15 min. Wait for the melted media to reach an approximate temperature of 55 °C before pouring it into sterile Petri plates (90 × 15 mm diameter). Store at 4 °C for up to 30 days [8].
6. *Trypticase Soy Agar (TSA)**, in g/L: 15.0 g of agar, 15.0 g of casein peptone, 5 g of sodium chloride, 5 g of soya peptone, and 1 L of distilled water. Sterilize by autoclaving at 121 °C for 15 min. Wait for the melted media to reach an approximate temperature of 55 °C before pouring it into sterile Petri plates (90 × 15 mm diameter). Store at 4 °C for up to 30 days [8].
7. *Seal agar 2%,* in g/L: 20 g of agar and 1 L of distilled water. Sterilize by autoclaving at 121 °C for 15 min.

*Commercially available.

### 2.2 Acid or Acidified Foods (pH $\leq$ 4.6)

1. *Themoacidurans broth (TAB)**, in g/L: 5.0 g of proteose peptone, 5.0 g of dextrose, 5.0 g of yeast extract, 0.2 g of manganese sulfate, and 1 L of distilled water. Sterilize by autoclaving at 121 °C for 15 min. For the solid version of the medium (*Thermoacidurans Agar, TAA*), add 15 g of agar per liter. Wait for the melted media to reach an approximate temperature of 55 °C before pouring it into sterile Petri plates (90 × 15 mm diameter). Store at 4 °C for up to 30 days [8].
2. *Potato dextrose agar (PDA)* acidified*, in g/L: 20.0 g of dextrose, 15.0 g of agar, 4 g of potato starch, and 1 L of distilled water. Cool and acidify to pH 3.5 with sterile 10% tartaric acid

solution and then mix thoroughly and pour into sterile Petri plates (90 × 15 mm diameter). Avoid heating the medium after adding tartaric acid to preserve its solidifying properties. Store at 4 °C for up to 30 days [8].

3. *All-purpose tween (APT) broth**, in g/L: 12.5 g of peptone, 10.0 g of dextrose, 7.5 g of yeast extract, 5.0 g of sodium citrate, 5.0 g of sodium chloride, 5.0 g of dipotassium hydrogen phosphate, 0.8 g of magnesium sulfate, 0.14 g of manganese (II) chloride, 0.04 g of ferrous sulfate, 0.001 g of thiamine hydrochloride, and 1 L of distilled water. For the solid version of the medium (*APT Agar*), add 13.5 g of agar per liter. After the dissolution of all ingredients, add 0.1 g of cycloheximide dissolved in a minimum amount of 40% ethanol and 0.2 mL Tween® 80. Sterilize by autoclaving at 121 °C for 15 min. Store at 4 °C for up to 30 days [8].
4. *De Man, Rogosa, and Sharpe (MRS) broth**, in g/L: 20.0 g of dextrose, 10.0 g of enzymatic digest of casein, 10.0 g of meat extract, 5.0 g of sodium acetate, 4.0 g of yeast extract, 2.0 g of dipotassium hydrogen phosphate, 2.0 g of triammonium citrate, 1.08 g of Tween® 80, 0.2 g of magnesium sulfate heptahydrate, 0.05 g of manganese sulfate tetrahydrate (or 0.04 g of manganese sulfate monohydrate), and 1 L distilled water. For the solid version of the medium (*MRS Agar*), add 14 g of agar per liter. Sterilize by autoclaving at 121 °C for 15 min. Store at 4 °C for up to 30 days [8].
5. *Orange serum broth (OSB)**, in g/L: 10.0 g of orange serum concentrate, 10.0 g of tryptone, 4.0 g of dextrose, 3.0 g of yeast extract, 2.5 g of dipotassium phosphate, and 1 L distilled water. For the solid version of the medium *(orange serum agar—OSA)*, add 18 g of agar per liter. Sterilize by autoclaving at 121 °C for 15 min. Store at 4 °C for up to 30 days [8].
6. *Dextrose tryptone agar (DTA)**, in g/L: 12.0 g of agar, 10.0 of tryptone, 5.0 g of dextrose, 0.04 g of bromocresol purple, and 1 L of distilled water. Dissolve the ingredients in distilled water by heating, if necessary. Sterilize by autoclaving at 121 °C for 15 min. Store at 4 °C for up to 30 days [8].
7. *ALI broth (optional)*, in g/L: 3.0 g of monopotassium phosphate, 2.0 g of soluble starch, 2.0 g of yeast extract, 1.0 g of dextrose, 0.5 g of magnesium sulfate monohydrate, 0.25 g of calcium chloride, 0.20 g of ammonium sulfate, and 1 L of distilled water. After mixing all ingredients, adjust pH to 3.5 with 1 N $H_2SO_4$ prior to autoclaving. Sterilize by autoclaving at 121 °C for 15 min. Store at 4 °C for up to 30 days [8].
8. *K Agar* (optional)*, in g/L: 15.0 g of agar, 5.0 g of peptone, 2.5 g of yeast extract, 1.0 g of dextrose, 1.0 g of Tween® 80, and 1 L of distilled water. Sterilize by autoclaving at

121 °C for 15 min. Cool to about 50 °C and adjust pH 3.7 ± 0.1 with 25% L-malic acid. Store at 4 °C for up to 30 days [8].

9. *Seal agar 2%*, in g/L: 20 g of agar and 1 L of distilled water. Sterilize by autoclaving at 121 °C for 15 min.

   *Commercially available.

### 2.3 Utensils, Reagents, and Equipments

1. *Bunsen burner.*
2. *Autoclave.*
3. *Deionized water.* Sterilize by autoclaving at 121 °C for 15 min. Store at 4 °C until use.
4. *Disinfectant agent.*
5. *Personal protective equipment and laminar flow workstation.*
6. *Permanent maker.*
7. *Scale.*
8. *Containment trays.*
9. *Incubators (25 °C, 30 °C, 35 °C, 45 °C, and 55 °C).*
10. *Water baths (45 °C, 80 °C, and 100 °C).*
11. *Bacteriological can opener*, for steel can foods.
12. *Box cutter*, for other types of packages.
13. *Sterile spoons, spatula, and tweezers.*
14. *Sterile containers*, such as Whirl-Pak® sterile bags or similar.
15. *Sterile pipettes (various volumes).*
16. *Microscope with 1000x magnification (phase-contrast recommended).*
17. *Slides, coverslips, loops, and immersion oil.*
18. *Gram staining reagents* (*see* **Note 5**).
19. *Spore staining reagents* (*see* **Note 6**).
20. *3% hydrogen peroxide.*

## 3 Methods

### 3.1 Pre-incubation

1. *Samples less than 30 days old.*

   Pre-incubate low-acid and/or acid/acidified food samples at temperatures of 30–35 °C and 25–30 °C, respectively, for 10 days to detect bacterial growth with gas formation. Swelling of the packaging and checking for possible microleaks indicate such growth.

2. *Samples manufactured more than 30 days ago.*

   Pre-incubation is not necessary.

3. *Altered samples.*

They should not be pre-incubated. Store in sturdy plastic bags and keep refrigerated until analysis.

### 3.2 Opening Samples Aseptically

Packaging should be cleaned with water and soap before disinfecting the food packaging with a disinfectant agent. In the case of flexible packaging (e.g., pouches and carton), use a scalpel or scissors flamed in the Bunsen burner flame to open the package. If the food is contained in a can, it should be flamed and opened using a bacteriological can opener. Combustible packaging should remain in contact with the disinfectant agent for 15 min and remove the excess with sterile cotton. It is crucial to highlight those packages presenting alterations, such as swelling, leakage, or corrosion, must be carefully handled. Swollen packages should be cooled before opening, which must be done cautiously to prevent the sudden release of gas. Moreover, such packages should not be subjected to alcohol flaming to avoid potential explosions.

### 3.3 Inoculation of Samples

After aseptically opening the packages, proceed with the inoculation of the tubes with 1–2 g of the samples in a laminar flow workstation, as shown in Table 1 (*see* **Note 1**). Incubate the tubes at appropriate temperature conditions (Table 1) to allow for microbial growth (*see* **Note 2**).

### 3.4 Isolation and Observation

All tubes described in Table 1 should be examined daily to detect any evidence of microbial growth, such as turbidity, gas production, and/or biofilm formation. In tubes where growth is detected, proceed with microbial characterization. In tubes with no evidence of growth, consider the product commercially sterile, pending a final microscopic evaluation of the subculture in each tube.

A portion of the content from each tested tube should be transferred to a slide and fixed using a Bunsen burner. Prior to observation, Gram staining, spore staining, or yeast staining may be performed. Additionally, certain characteristics, such as motility may be observed.

### 3.5 Characterization of Microbial Contamination and Interpretation of Results

#### 3.5.1 Test of Commercial Sterility for Low-Acid Foods

1. *DTB*

Once microbial growth is detected in the DTB tubes, characterize the culture (*see* **Note 4**) using the spread plate technique with 1 mL of inoculum dispensed onto four NAMn agar plates. Incubate two plates at 55 °C for 4 days and the remaining two plates at 30–35 °C for 7–10 days (Fig. 1).

Following incubation, select distinct typical colonies on NAMn agar plates and conduct Gram (*see* **Note 5**) and spore (*see* **Note 6**) staining for microscopic analysis. For plates incubated at 30–35 °C, the presence of Gram-positive rods exhibiting spore formation indicates the existence of aerobic spore-forming mesophilic or facultative thermophilic bacteria.

**Table 1**
**Commercial sterility testing conditions for low-acid and acid or acidified foods [8–10].**

| Type of sample | Media | Quantity | Temperature/incubation period | Cause of spoilage |
|---|---|---|---|---|
| Low-acid food (*see* Fig. 1) | DTB | Two tubes | 30–35 °C/7–10 days | Mesophilic or facultative thermophilic aerobic spore-forming bacteria |
| | CMM or PE-2 with seal agar 2% (*see* **Note 3**) | | 30–35 °C/7–10 days | Mesophilic anaerobic spore-forming bacteria |
| | DTB | | 55 °C/4 days | Facultative or obligate thermophilic aerobic spore-forming bacteria |
| | CF or CMM or PE-2 with seal agar 2% (*see* **Note 3**) | | 55 °C/4 days | Thermophilic anaerobic spore-forming bacteria |
| | LVA | Two plates | 30–35 °C/7–10 days | Mesophilic anaerobic spore-forming bacteria |
| | LVA | | 55 °C/4 days | Thermophilic anaerobic spore-forming bacteria |
| | NAMn | | 30–35 °C/7–10 days | Mesophilic or facultative thermophilic aerobic spore-forming bacteria |
| | NAMn | | 55 °C/4 days | Obligate thermophilic aerobic spore-forming bacteria |
| | TSA (Control) | | 30–35 °C/7 days (aerobic conditions) | Mesophilic aerobic spore-forming bacteria |
| | TSA (Control) | | 30–35 °C/7 days (anaerobic conditions) | Mesophilic anaerobic spore-forming bacteria |
| Acid or acidified foods (*see* Fig. 2) | TAB with seal agar 2% | Two tubes | 30–35 °C/7 days | Aciduric mesophilic aerobic and anaerobic bacteria |
| Acid or acidified foods (*see* Fig. 2) | TAB | Two tubes | 30–35 °C/7 days | Aciduric mesophilic aerobic and facultative anaerobic bacteria |
| | TAB | | 55 °C/7 days | Aciduric thermophilic aerobic bacteria |
| | PDA (acidified) | | 25–30 °C/7 days | Molds and yeast |
| | APT, MRS, or OSB (optional, for detection of lactic acid bacteria) | | 30 °C/7 days | Lactic acid bacteria |
| | ALI or K Agar (optional, for detection of *Alicyclobacillus*) | | 43 °C/7 days | *Alicyclobacillus* |
| | TAA (Control) | Two plates | 30–35 °C (Aerobe)/2–5 days | Aciduric mesophilic aerobic bacteria |
| | TAA (Control) | | 30–35 °C (Anaerobe)/ 2–5 days | Aciduric mesophilic anaerobic bacteria |

Transfer a loopful of culture from the plates initially incubated at 55 °C for 4 days into a tube containing 1 mL of distilled water and subject the sample to thermal shock by boiling for 10 min, followed by immediate cooling in an ice bath. Inoculate a loopful of the treated suspension onto two NAMn agar plates, incubating one plate at 30–35 °C for 4 days and the other at 55 °C. Growth observed at 30–35 °C indicates the presence of facultative thermophilic aerobic spore-forming bacteria, whereas growth at 55 °C suggests obligate aerobic spore-forming bacteria.

2. *CMM or PE-2*

Once microbial growth is detected in the CMM or PE-2 tubes, characterize the culture (*see* **Notes 3** and **4**) using the pour plate technique with 1 mL of inoculum dispensed onto eight LVA agar plates. Incubate four plates at 55 °C for 4 days (two aerobically and two anaerobically) and the remaining four plates at 30–35 °C for 7–10 days (similarly divided into aerobic and anaerobic conditions) (Fig. 1).

Following incubation, select morphologically distinct colonies from the NAMn agar plates and perform Gram staining (*see* **Note 5**) and spore staining (*see* **Note 6**) for microscopic analysis. Cultures incubated at 30–35 °C and 55 °C are classified as mesophilic anaerobic spore-forming bacteria and thermophilic anaerobic spore-forming bacteria, respectively, based on microscopic observations.

#### *3.5.2 Test of Commercial Sterility for Acid or Acidified Foods*

1. *TAB with seal agar 2%*

Upon detection of microbial growth in the TAB tubes sealed with 2% agar, characterize the culture using the pour plate technique by transferring 1 mL of inoculum onto TAA agar plates. Incubate two plates under aerobic conditions and two plates under anaerobic conditions at 30–35 °C for 2–5 days (Fig. 2) to determine whether the contamination is facultative or obligate anaerobic. Following incubation, select morphologically distinct colonies from the TAA agar plates and perform Gram staining (*see* **Note 5**) and spore staining (*see* **Note 6**) for microscopic analysis. During observation, confirm the presence of rod-shaped bacteria exhibiting spore formation, gas production, and a butyric odor. This observation is likely indicative of thermal underprocessing.

2. *TAB*

Upon detection of microbial growth in the TAB tubes incubated at 55 °C and 30–35 °C, characterize the microorganisms using the spread plate technique by dispensing 1 mL of inoculum onto TAA agar plates for each temperature range. Incubate one plate at 30–35 °C and one plate at 55 °C (from each tube) for 2–5 days (Fig. 2) to determine whether the contamination is

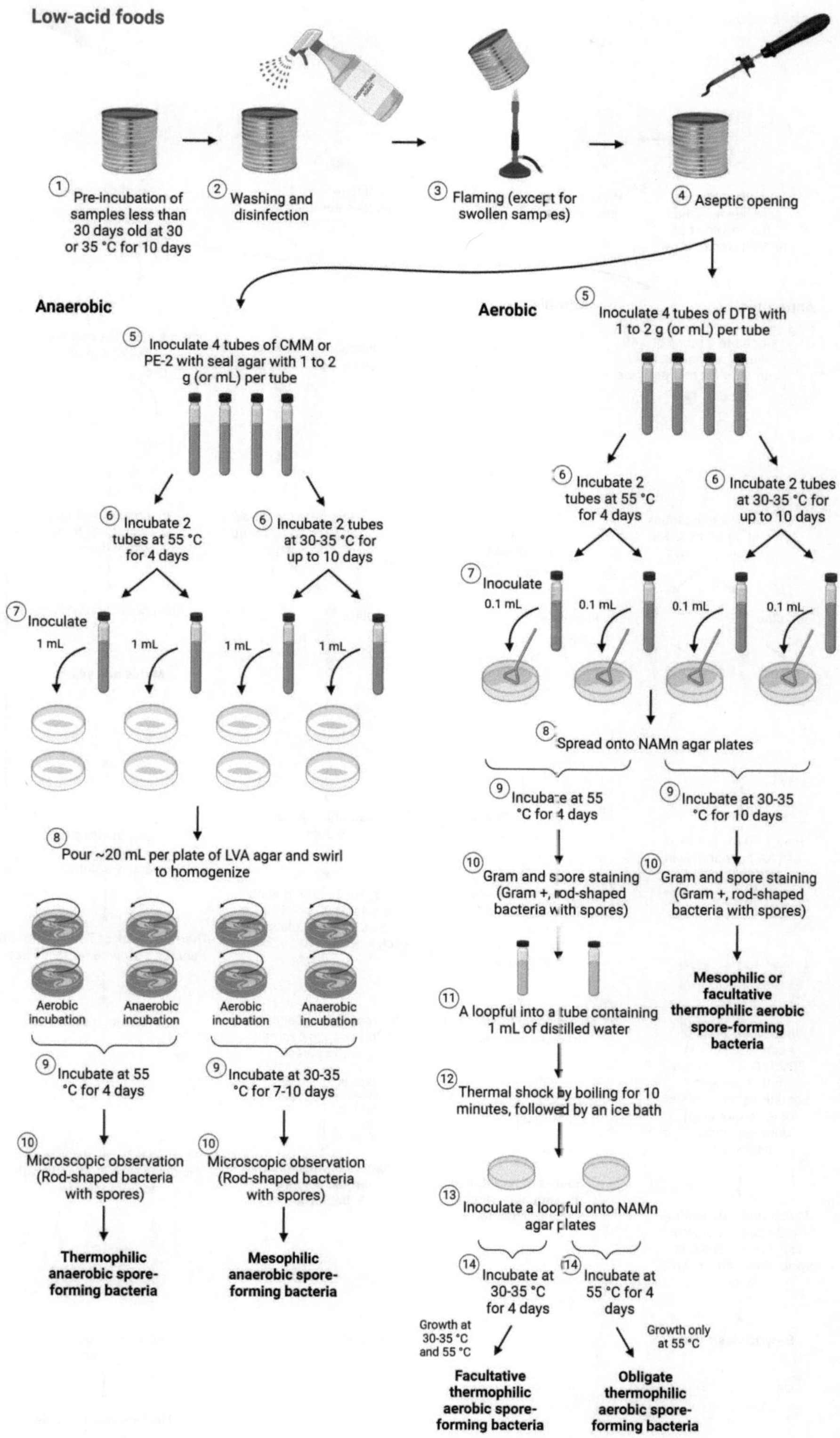


**Fig. 1** Flowchart for commercial sterility testing of low-acid foods by the APHA:2015 method. (Modified and adapted from: [8–10])

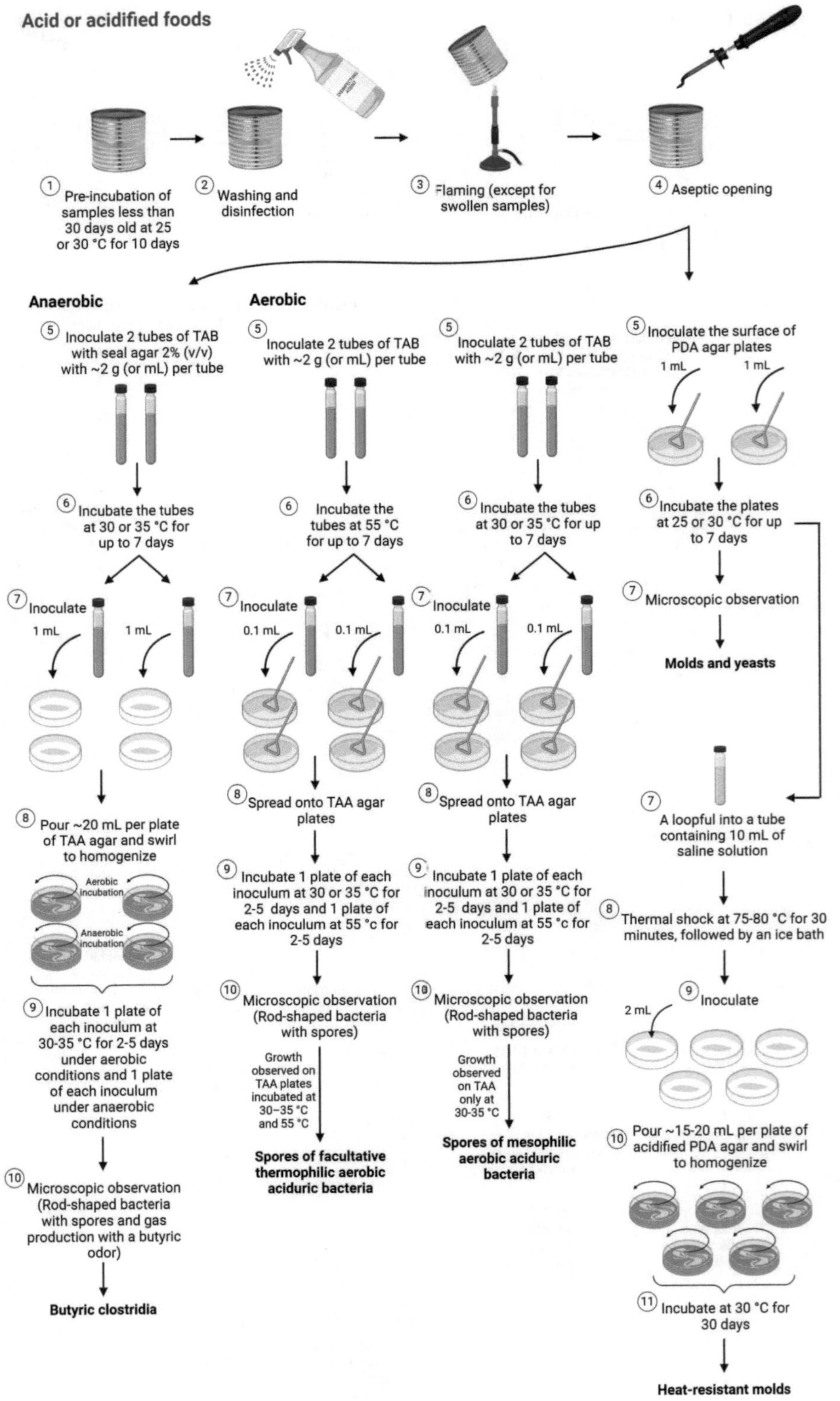


**Fig. 2** Flowchart for commercial sterility testing of acid or acidified foods by the APHA:2015 method. (Modified and adapted from: [8–10])

mesophilic or facultative thermophilic aerobic spore-forming bacteria. Following incubation, select morphologically distinct colonies from the TAA agar plates and perform Gram staining (*see* **Note 5**) and spore staining (*see* **Note 6**) for microscopic analysis. During observation, confirm the presence of rod-shaped bacteria exhibiting spore formation. If growth is observed on TAA plates incubated at both 30–35 °C and 55 °C, this confirms the presence of spores from facultative thermophilic aerobic aciduric bacteria. If growth is observed exclusively on TAA plates incubated at 30–35 °C, this confirms the presence of spores from mesophilic aerobic aciduric bacteria. This observation is likely indicative of thermal underprocessing. The exclusive growth of thermophilic aerobic aciduric bacteria indicates deterioration caused by slow cooling and/or excessively high storage temperatures.

3. *PDA (Acidified)*

Following incubation, select morphologically distinct colonies from the acidified PDA plates incubated at 25–30 °C for up to 7 days. The observed growth may indicate the presence of molds and yeasts, suggesting deterioration due to leakage. Transfer a loopful of a colony from the acidified PDA agar plate into a tube containing 10 mL of saline solution, and subject it to thermal shock at 75–80 °C for 30 min, followed by immersion in an ice bath. Subsequently, distribute 2 mL of the solution across five acidified PDA agar plates and incubate at 30 °C for up to 30 days. Growth observed on the acidified PDA agar plates confirms the presence of heat-resistant molds in the original product, indicating deterioration due to thermal underprocessing.

4. *APT, MRS, or OSB (optional)*

Upon detection of microbial growth in TAB tubes with or without a 2% agar seal, transfer a loopful from each tube to APT, MRS, or OSB agar plates. Incubate these plates at 30 °C for 2–5 days under microaerophilic conditions (*see* **Note 7**). Select at least five colonies for catalase testing (*see* **Note 8**) and Gram staining (*see* **Note 5**). The observation of Gram-positive, catalase-negative bacteria exhibiting rod-shaped, coccal, or coccobacillary morphology confirms contamination of the original product with lactic acid bacteria. This observation is likely indicative of thermal underprocessing or leakage.

5. *ALI or K Agar (this step is optional and should be performed if Alicyclobacillus spoilage is suspected)*

Upon detection of microbial growth in TAB tubes with or without a 2% agar seal, transfer a loopful from each tube to a K Agar plate and incubate at 43 °C for 2–5 days. Following incubation, select morphologically distinct colonies from the K Agar plates and perform Gram staining (*see* **Note 5**) and spore staining (*see* **Note 6**) for microscopic analysis. In addition to observing rod-shaped, spore-forming bacteria, assess the presence of off-flavor odors, which are likely indicative of guaiacol production by *Alicyclobacillus* species. This observation is likely indicative of thermal underprocessing.

## 4 Notes

1. A sample of approximately 50 g or 50 mL should be transferred into each of two sterile collection bags to be stored as a retained sample.
2. If the tested food is not stored at temperatures above 37 °C, the tubes designated for incubation at 55 °C may be excluded from the assessment.
3. When handling tubes incubated at 30–35 °C, exercise caution due to the potential presence of *C. botulinum* and its associated toxins. Proper safety protocols must be strictly adhered to mitigate the risks of exposure and contamination.
4. In cases of uncertainty regarding microbial growth, proceed to the subculture step. If no growth is observed on the plates, the original tube should be interpreted as yielding a negative result.
5. Gram staining procedure: (1) Prepare a bacterial smear on a glass slide and allow it to air dry. (2) Flood the smear with crystal violet stain for 1 min and then rinse gently with water. (3) Apply Gram's iodine solution for 1 min, followed by a gentle water rinse. (4) Decolorize using alcohol or acetone for 10–15 s and then rinse immediately with water. (5) Counterstain with safranin for 1 min, followed by a final water rinse. (6) Allow the slide to air dry and then examine under a microscope. Gram-positive bacteria will appear purple, while Gram-negative bacteria will appear pink.
6. Spore staining procedure: (1) Stain with malachite green and steam for 5–10 min and then rinse with water. (2) Stain with safranin for 1 min and then rinse with water. (3) Allow the slide to air dry and then examine under a microscope. Spores will appear green, while vegetative cells will appear pink.
7. Agar overlay procedure: (1) Pour and solidify a base agar layer. (2) Spread bacteria on the base layer. (3) Pour a thin layer (~ 5 mL) of molten agar over the bacteria. (4) Let it solidify. (5) Incubate and observe growth in the reduced oxygen environment.
8. Catalase test: Add 1–2 mL of 3% hydrogen peroxide to the bacteria. Look for bubbles: Bubbles—Catalase-positive. No bubbles—Catalase-negative.

## Acknowledgments

Leonardo do Prado-Silva acknowledges the financial support of the "Fundação de Amparo à Pesquisa do Estado de São Paulo" (FAPESP): Grant #2023/18197-7. Caroline Heckler acknowledges the financial support of the "Fundação de Amparo à Pesquisa do Estado de São Paulo" (FAPESP): Grant #2020/05580-9.

### References

1. Vannini L, Siroli L (2025) Canned and heat preserved foods. In: Bevilacqua A, Corbo MR, Sinigaglia M (eds) The microbiological quality of food, 2nd edn. Elsevier, pp 261–271
2. IFTPS (2014) Guidelines for conducting thermal processing studies, Guelph
3. Carlin F (2011) Origin of bacterial spores contaminating foods. Food Microbiol 28:177–182. https://doi.org/10.1016/j.fm.2010.07.008
4. International Commission on Microbi (2018) Microorganisms in foods 7. Springer International Publishing, Cham
5. Fleischman GJ (2024) Food sterilization. In: Encyclopedia of food safety. Elsevier, pp 432–444
6. Rolfe C, Daryaei H (2020) Intrinsic and extrinsic factors affecting microbial growth in food systems. In: Demirci A, Feng H, Krishnamurthy K (eds) Food safety engineering. Springer, pp 3–24
7. Snyder AB, Worobo RW (2018) The incidence and impact of microbial spoilage in the production of fruit and vegetable juices as reported by juice manufacturers. Food Control 85:144–150. https://doi.org/10.1016/j.foodcont.2017.09.025
8. da Silva N, Junqueira VCA, Ferraz N, de Silveira A et al (2021) Manual de métodos de análise microbiológica de alimentos e água, 6th edn. Blucher, São Paulo
9. Elliott PH, Kataoka A (2015) 61. Canned foods—tests for commercial sterility. In: Compendium of methods for the microbiological examination of foods. American Public Health Association. https://doi.org/10.2105/MBEF.0222.066
10. Parkinson NG, Francis K (2015) 62. Canned foods—tests for cause of spoilage. In: Compendium of methods for the microbiological examination of foods. American Public Health Association. https://doi.org/10.2105/MBEF.0222.067

# Chapter 5

# Preparation of Suspensions of Bacterial Vegetative Cells


**Jaqueline Sousa Correia and Matheus Péricles Silva Láscaris**



## Abstract

Pathogenic and spoilage bacteria compromise the quality and safety of food. In contrast, beneficial microorganisms are essential as they improve sensory characteristics and inhibit the growth of pathogens. These aspects highlight the need to evaluate and control microbial behavior. In this regard, the use of bacterial vegetative cell suspensions in experimental studies is important. Therefore, the objective of this chapter is to detail the processes for obtaining these suspensions.



**Key words** Bacterial suspensions, Food safety, Vegetative cells


## 1 Introduction

Food can contain both beneficial microorganisms [1, 2] and pathogens [3, 4], which affect its quality, safety, and shelf life. Pathogens such as *Salmonella*, *Escherichia coli*, and *Listeria* cause diseases, spoilage, and changes in organoleptic characteristics, reducing the nutritional value and compromising food quality and safety [5–7]. Fungi [8] also contribute to food spoilage by producing mycotoxins [9] that affect product quality and consumer safety. On the other hand, beneficial microorganisms, such as lactic acid bacteria and yeasts, play an important role in improving the sensory characteristics and inhibiting pathogen growth [10–13].

Understanding and controlling microbial behavior is essential for ensuring food quality and safety. In this context, bacterial vegetative cell suspensions, which are active forms of growth and metabolism [14], are important tools for this understanding and are fundamental in microbial inactivation studies [15, 16], challenge tests [17], and other experimental applications [18–21].

For this, it is necessary to use pure bacterial solutions to reduce contamination from cell residues and secondary metabolic byproducts, which are inherent to cell growth and may interfere with various metabolic processes in microorganisms [16–18]. This


Anderson S. Sant'Ana (ed.), *Heat Resistant Microorganisms in Food Safety and Spoilage*, Methods and Protocols in Food Science, https://doi.org/10.1007/978-1-0716-5280-0_5,

chapter aims to detail and assist in the processes for obtaining bacterial vegetative cell suspensions.

## 2 Materials

### 2.1 Media

1. *Trypticase Soy Broth (TSB):* 17 g Casein peptone (enzymatic digestion), 3 g soy peptone (enzymatic digestion), 5 g sodium chloride (NaCl), 2.5 g dipotassium phosphate (K2HPO4), 2.5 g dextrose, and 1000 mL purified water. Adjust the final pH to 7.3 ± 0.2 using 1 N sodium hydroxide (NaOH) solution while stirring. Autoclave at 121 °C for 15 min.
2. *Trypticase Soy Agar (TSA):* 15 g Casein peptone (enzymatic digestion), 3 g soy peptone (enzymatic digestion), 5 g sodium chloride (NaCl), 15 g agar, and 1000 mL purified water. Adjust the final pH to 7.3 ± 0.2 with 1 N NaOH solution while stirring. Autoclave at 121 °C for 15 min.

### 2.2 Solutions and Reagents

1. *Phosphate Saline Solution:* Dissolve 7.650 g sodium chloride (NaCl), 0.724 g disodium phosphate (Na2HPO4), and 0.210 g monopotassium phosphate (KH2PO4) in 1000 mL purified water. Stir and adjust the final pH to 7.4 ± 0.2 with 1 N sodium hydroxide (NaOH) solution. Autoclave at 121 °C for 15 min.
2. *0.85% Saline Solution:* Dissolve 8.5 g sodium chloride (NaCl) in 1000 mL purified water. Autoclave at 121 °C for 15 min.

## 3 Methods

### 3.1 Reactivation of a Cryopreserved Culture

1. Transfer 10 μL of the cryopreserved culture to a tube containing 5 mL of TSB broth.
2. Mix using a vortex and incubate under the conditions specified for the microorganism (time/temperature).
3. Streak the culture onto a TSA agar plate to obtain isolated colonies. Incubate the plate under the recommended conditions for the microorganism.

### 3.2 Propagation to Stabilize the Basal Phenotype

1. Select one to three isolated colonies and transfer them into 10 mL of TSB broth. Mix using a vortex and incubate under the recommended conditions for the microorganism.
2. After the incubation period, add 100 μL of the pre-culture to 10 mL of TSB broth and incubate again under the recommended conditions for the microorganism.

**3.3 Increase in Biomass Production**

4. Add 100 μL of the culture to 100 mL of TSB broth and incubate with constant agitation (100–200 rpm) under the conditions specified for the microorganism.

**3.4 Biomass Collection**

1. After the incubation period, centrifuge the culture according to the microorganism's specific conditions (g-force or rpm, time, and temperature) (*see* **Notes 1** and **2**).
2. Then, carefully remove the supernatant to avoid damaging the pellet.
3. Add 100 mL of phosphate-buffered saline (PBS) or 0.85% saline solution and vortex to dissolve the pellet.
4. Repeat the washing step (centrifugation and resuspension) two more times.
5. Finally, add 50 mL of the solution used (PBS or 0.85% saline solution), obtaining the vegetative cell suspension.

**3.5 Verification and Adjustment of Cell Concentration**

1. Transfer 1 mL of the suspension into a spectrophotometer cuvette.
2. Measure the optical density of the cell suspension at 630 nm, ensuring that the reading is between 0.1 and 0.9, according to Beer's law (*see* **Note 3**).
3. Plate 0.1 mL of the solution on TSA agar (perform dilutions if necessary) and incubate under the recommended conditions for the microorganism.
4. Count the colonies on the plate and calculate the concentration of the cell suspension in CFU/mL.

**3.6 Construction of the Calibration Curve**

1. To construct the calibration curve, select five or more absorbance points and plate on TSA agar as described above. The absorbance can be adjusted by adding the cell suspension or stock solutions.

**Table 1**
**Example of experimental data for the calibration curve**

| Absorbance | CFU/mL | LogCFU/mL |
|---|---|---|
| 0.100 | 4.50E+03 | 3.65 |
| 0.300 | 2.24E+04 | 4.35 |
| 0.500 | 2.79E+05 | 5.44 |
| 0.700 | 3.55E+06 | 6.55 |
| 0.900 | 4.50E+07 | 7.65 |

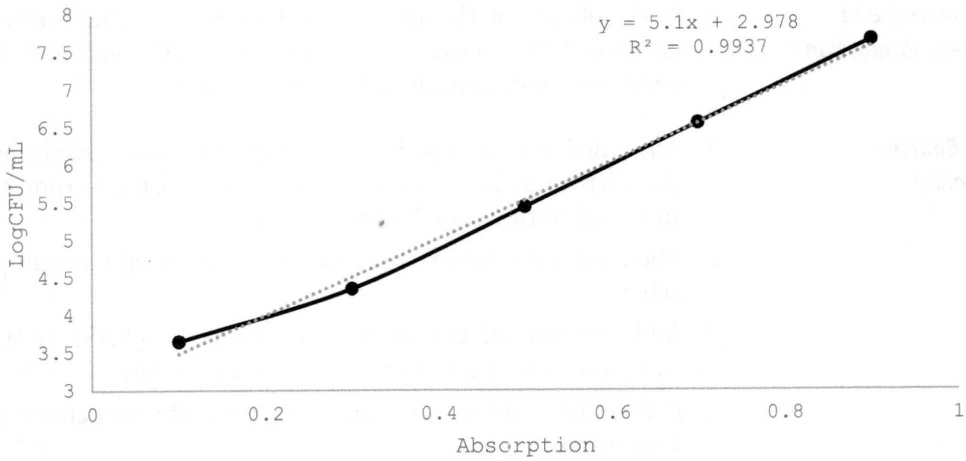


**Fig. 1** Linear fit with the equation of the line from the experimental data

2. Using software, construct a linear regression curve using the data obtained (Absorbance x LogCFU/mL) (Table 1 and Fig. 1).

3. Obtain the equation of the line from the linear regression curve.

4. Using the equation of the line, it is possible to determine the concentration of vegetative cells within the range established in the experiment.

## 4 Notes

1. Centrifugation parameters for vegetative bacterial cells are shown in Table 2.
2. If necessary, convert g-force to rpm (Eq. 1):

$$G - force = 1,12 * R * \left(\frac{RPM^2}{1.000}\right) \quad (1)$$

3. The Beer–Lambert law states that absorbance is directly proportional to the concentration of a solution, provided the experimental conditions are appropriate. Absorbance values below 0.1 indicate low absorption, leading to imprecise readings due to background noise or detector limitations. Values above 0.9 indicate high absorption, which reduces the amount of transmitted light to minimal levels, compromising detection accuracy (Eq. 2).

$$A = \varepsilon b \quad (2)$$

**Table 2**
**Centrifugation parameters for microorganisms of food importance**

| Microorganism | G-force | rpm | Time (min) | Temperature (°C) | Reference |
|---|---|---|---|---|---|
| Molds and yeasts | 6.500–10.000 | 5.000–6.000 | 5–10 | 4 | [22–25] |
| *Enterobacter* spp. | 340–14.000 | – | 10–30 | 4 | [26–28] |
| *Escherichia coli* | 6.000–8.000 | 4.000 | 2–10 | 4 | [29–32] |
| *Staphylococcus aureus* | 1.912 | 4.000–4.500 | 10–15 | 4 | [33–35] |
| *Bacillus* spp. | 800–15.000 | – | 3–10 | 4 | [36–38] |
| *Clostridium* spp. | 4.000–15.000 | 10.000 | 10–30 | 4 | [39–41] |
| *Campylobacter* spp. | 3.300–10.000 | – | 6–30 | 4 | [42, 43] |
| *Cronobacter* spp. | 10.000 | – | 10 | 4 | [44] |
| *Vibrio* spp. | 4.500 | 1.000 | 5–30 | 4 | [45, 46] |
| *Listeria monocytogenes* | 2.810 | 4.700 | 10–15 | 4 | [47, 48] |
| *Yersinia* spp. | 2.236 | – | 20 | 4 | [49] |
| *Salmonella* spp. | 4.000–11.088 | – | 5–10 | 4 | [50, 51] |
| Lactic acid bacteria | 4.000–5.000 | – | 5–10 | 4 | [52, 53] |

## References


1. Adesulu-Dahunsi A, Dahunsi S, Ajayeoba T (2022) Co-occurrence of Lactobacillus species during fermentation of African indigenous foods: impact on food safety and shelf-life extension. Front Microbiol 13:684730. https://doi.org/10.3389/fmicb.2022.684730
2. Lahiri D, Nag M, Sarkar T, Ray R, Shariati M, Rebezov M, Bangar S, Lorenzo J, Domínguez R (2022) Lactic Acid Bacteria (LAB): autochthonous and probiotic microbes for meat preservation and fortification. Foods 11:2792. https://doi.org/10.3390/foods11182792
3. McMeekin T, Brown J, Krist K, Miles D, Neumeyer K, Nichols D, Olley J, Presser K, Ratkowsky D, Ross T, Salter M, Soontranon S (1997) Quantitative microbiology: a basis for food safety. Emerg Infect Dis 3:541–549
4. Brackett R (1992) Shelf stability and safety of fresh produce as influenced by sanitation and disinfection. J Food Prot 55(10):808–814. https://doi.org/10.4315/0362-028X-55.10.808
5. Cooley M, Quiñones B, Oryang D, Mandrell R, Gorski L (2014) Prevalence of Shiga toxin producing Escherichia coli, Salmonella enterica, and Listeria monocytogenes at public access watershed sites in a California central coast agricultural region. Front Cell Infect Microbiol 4. https://doi.org/10.3389/fcimb.2014.00030
6. Duc H, Son H, Yi H, Sato J, Ngan P, Masuda Y, Honjoh K, Miyamoto T (2020) Isolation, characterization and application of a polyvalent phage capable of controlling Salmonella and Escherichia coli O157:H7 in different food matrices. Food Res Int 131:108977. https://doi.org/10.1016/j.foodres.2020.108977
7. Sheng L, Wang L (2020) The microbial safety of fish and fish products: recent advances in understanding its significance, contamination sources, and control strategies. Compr Rev Food Sci Food Saf 20(1):738–786. https://doi.org/10.1111/1541-4337.12671
8. Benedict K, Chiller T, Mody R (2016) Invasive fungal infections acquired from contaminated food or nutritional supplements: a review of the literature. Foodborne Pathog Dis 13(7):

343–349. https://doi.org/10.1089/fpd.2015.2108

9. Bennett JW, Klich M (2003) Mycotoxins. Clin Microbiol Rev 16(3):497–516. https://doi.org/10.1128/CMR.16.3.497-516.2003
10. Sharma R, Garg P, Kumar P, Bhatia S, Kulshrestha S (2020) Microbial fermentation and its role in quality improvement of fermented foods. Fermentation 6:106. https://doi.org/10.3390/fermentation6040106
11. Adesulu-Dahunsi A, Dahunsi S, Olayanju A (2020) Synergistic microbial interactions between lactic acid bacteria and yeasts during production of Nigerian indigenous fermented foods and beverages. Food Control 110:106963. https://doi.org/10.1016/j.foodcont.2019.106963
12. Smid E, Lacroix C (2013) Microbe-microbe interactions in mixed culture food fermentations. Curr Opin Biotechnol 24(2):148–154. https://doi.org/10.1016/j.copbio.2012.11.007
13. Arena M, Capozzi V, Russo P, Drider D, Spano G, Fiocco D (2018) Immunobiosis and probiosis: antimicrobial activity of lactic acid bacteria with a focus on their antiviral and antifungal properties. Appl Microbiol Biotechnol 102:9949–9958. https://doi.org/10.1007/s00253-018-9403-9
14. Madigan MT, Martinko JM, Parker J (2018) Brock biology of microorganisms, 15th edn. Pearson
15. Cronin U, Wilkinson M (2008) Physiological response of Bacillus cereus vegetative cells to simulated food processing treatments. J Food Prot 71(11):2168–2176. https://doi.org/10.4315/0362-028X-71.11.2168
16. Liossi LL, Heckler C, Sant'Ana AS (2024) High-pressure carbon dioxide (HPCD): impact on the quality of fruit juices and inactivation of spores and enzymes. Food Res Int 198:115316. https://doi.org/10.1016/j.foodres.2024.115316
17. Augustin J, Bergis H, Midelet-Bourdin G, Cornu M, Couvert O, Denis C, Huchet V, Lemonnier S, Pinon A, Vialette M, Zuliani V, Stahl V (2011) Design of challenge testing experiments to assess the variability of Listeria monocytogenes growth in foods. Food Microbiol 28(4):746–754. https://doi.org/10.1016/j.fm.2010.05.028
18. Hussain M, Tango C, Oh D (2019) Inactivation kinetics of slightly acidic electrolyzed water combined with benzalkonium chloride and mild heat treatment on vegetative cells, spores, and biofilms of Bacillus cereus. Food Res Int 116:157–167. https://doi.org/10.1016/j.foodres.2018.08.003
19. Kennedy D, Cronin U, Wilkinson M (2011) Responses of Escherichia coli, listeria monocytogenes, and Staphylococcus aureus to simulated food processing treatments, determined using fluorescence-activated cell sorting and plate counting. Appl Environ Microbiol 77:4657–4668. https://doi.org/10.1128/AEM.00323-11
20. Furukawa S, Noma S, Shimoda M, Hayakawa I (2002) Effect of initial concentration of bacterial suspensions on their inactivation by high hydrostatic pressure. Int J Food Sci Technol 37:573–577. https://doi.org/10.1046/J.1365-2621.2002.00586.X
21. Cebrián G, Condón S, Mañas P (2017) Physiology of the inactivation of vegetative bacteria by thermal treatments: mode of action, influence of environmental factors and inactivation kinetics. Foods 6. https://doi.org/10.3390/foods6120107
22. Sena ARD, Valasques Júnior GL, Assis SA (2012) Application of Doehlert experimental design in the optimization of experimental variables for the Pseudozyma sp. (CCMB 306) and Pseudozyma sp. (CCMB 300) cell lysis. Food Sci Technol 32:761–767
23. Liu X, Jia B, Sun X, Ai J, Wang L, Wang C et al (2015) Effect of initial pH on growth characteristics and fermentation properties of Saccharomyces cerevisiae. J Food Sci 80(4):M800–M808
24. Velásquez A, Arias R, Toneatti M (2012) Effect of the type of substrate on the chemical composition and productivity of a protein concentrate of yeast origin. Ciencia e investigación agraria: revista latinoamericana de ciencias de la agricultura 39(3):425–434
25. Pereyra CM, Gil S, Cristofolini A, Bonci M, Makita M, Monge MDP et al (2018) The production of yeast cell wall using an agroindustrial waste influences the wall thickness and is implicated on the aflatoxin B1 adsorption process. Food Res Int 111:306–313
26. Bojórquez C, Frías-Espericueta MG, Gómez-Gil B, Voltolina D (2019) Biosorption of cadmium and lead using suspended and immobilized Enterobacter cloacae at different pH. Rev Int Contam Ambient 35(1):259–264
27. Feria-Cáceres PF, Penagos-Velez L, Moreno-Herrera CX (2022) Tolerance and cadmium (cd) immobilization by native bacteria isolated in cocoa soils with increased metal content. Microbiol Res 13(3):556–573
28. Pasteran F, Gonzalez LJ, Albornoz E, Bahr G, Vila AJ, Corso A (2016) Triton Hodge test:

improved protocol for modified Hodge test for enhanced detection of NDM and other carbapenemase producers. J Clin Microbiol 54(3): 640–649
29. Chen J, Alcaine SD, Jiang Z, Rotello VM, Nugen SR (2015) Detection of Escherichia coli in drinking water using T7 bacteriophage-conjugated magnetic probe. Anal Chem 87(17):8977–8984
30. Gyawali R, Adkins A, Minor RC, Ibrahim SA (2014) Behavior and changes in cell morphology of Escherichia coli O157: H7 in liquid medium and skim milk in the presence of caffeine. CyTA-J Food 12(3):235–241
31. Borja GM, Meza Mora E, Barrón B, Gosset G, Ramírez OT, Lara AR (2012) Engineering Escherichia coli to increase plasmid DNA production in high cell-density cultivations in batch mode. Microb Cell Factories 11:1–9
32. Li F, Xiong XS, Yang YY, Wang JJ, Wang MM, Tang JW et al (2021) Effects of NaCl concentrations on growth patterns, phenotypes associated with virulence, and energy metabolism in Escherichia coli BW25113. Front Microbiol 12:705326
33. Lee S, Choi KH, Yoon Y (2014) Effect of NaCl on biofilm formation of the isolate from Staphylococcus aureus outbreak linked to ham. Korean J Food Sci Anim Resour 34(2):257
34. Amaya Arbelaez MI, de Paula e Silva ACA, Navegante G, Valente V, Barbugli PA, Vergani CE (2021) Proto-oncogenes and cell cycle gene expression in normal and neoplastic oral epithelial cells stimulated with soluble factors from single and dual biofilms of Candida albicans and Staphylococcus aureus. Front Cell Infect Microbiol 11:627043
35. Oliva MDLM, Carezzano ME, Gallucci MN, Freytes SA, Zygadlo JA, Demo MS (2015) Growth inhibition and morphological alterations of Staphylococcus aureus caused by the essential oil of Aloysia triphylla. Bol Latinoam Caribe Plantas Med Aromat 14(2):83–91
36. Jiang R, Wu X, Xiao Y, Kong D, Li Y, Wang H (2021) Tween 20 regulate the function and structure of transmembrane proteins of Bacillus cereus: promoting transmembrane transport of fluoranthene. J Hazard Mater 403: 123707
37. de León L, Beltrán B, Moujir L (2005) Antimicrobial activity of 6-oxophenolic triterpenoids. Mode of action against Bacillus subtilis. Planta Med 71(04):313–319
38. De León L, Moujir L (2008) Activity and mechanism of the action of zeylasterone against Bacillus subtilis. J Appl Microbiol 104(5):1266–1274
39. Goldstein J, Morris WE, Loidl CF, Tironi-Farinatti C, McClane BA, Uzal FA, Fernandez Miyakawa ME (2009) Clostridium perfringens epsilon toxin increases the small intestinal permeability in mice and rats. PLoS One 4(9): e7065
40. Mah JH, Kang DH, Tang J (2009) Comparison of viability and heat resistance of Clostridium sporogenes stored at different temperatures. J Food Sci 74(1):M23–M27
41. Lavilla M, Marzo I, De Luis R, Perez MD, Calvo M, Sánchez L (2010) Detection of Clostridium tyrobutyricum spores using polyclonal antibodies and flow cytometry. J Appl Microbiol 108(2):488–498
42. Rodrigues RC, Pocheron AL, Hernould M, Haddad N, Tresse O, Cappelier JM (2015) Description of Campylobacter jejuni Bf, an atypical aero-tolerant strain. Gut Pathog 7:1–12
43. Biswas D, Fernando U, Reiman C, Willson P, Potter A, Allan B (2006) Effect of cytolethal distending toxin of Campylobacter jejuni on adhesion and internalization in cultured cells and in colonization of the chicken gut. Avian Dis 50(4):586–593
44. Chen Q, Li Y, Tao T, Bie X, Lu F, Lu Z (2017) Development and application of a sensitive, rapid, and reliable immunomagnetic separation-PCR detection method for Cronobacter spp. J Dairy Sci 100(2):961–969
45. Srinivasan P, Ramasamy P (2017) Morphological characterization and biocontrol effects of Vibrio vulnificus phages against Vibriosis in the shrimp aquaculture environment. Microb Pathog 111:472–480
46. González Y, Camarena L, Dreyfus G (2015) Induction of the lateral flagellar system of Vibrio shilonii is an early event after inhibition of the sodium ion flux in the polar flagellum. Can J Microbiol 61(3):183–191
47. Sant'Ana AS, Barbosa MS, Destro MT, Landgraf M, Franco BD (2012) Growth potential of Salmonella spp. and Listeria monocytogenes in nine types of ready-to-eat vegetables stored at variable temperature conditions during shelf-life. Int J Food Microbiol 157(1): 52–58
48. Martins M, Custódio R, Camejo A, Almeida MT, Cabanes D, Sousa S (2012) Listeria monocytogenes triggers the cell surface expression of Gp96 protein and interacts with its N terminus to support cellular infection. J Biol Chem 287(51):43083–43093
49. Lim JS, Ha JW (2022) Growth temperature-induced changes in resistance of Listeria

monocytogenes and Yersinia enterocolitica to X-ray irradiation. Food Microbiol 105:104029

50. Graça-Lopes G, Graça G, Barahona S, Moreira RN, Arraiano CM, Gonçalves LG (2019) NMR-metabolomics shows that BolA is an important modulator of Salmonella typhimurium metabolic processes under virulence conditions. Meta 9(11):243

51. Mattiello SP, Barth VC Jr, Scaria J, Ferreira CAS, Oliveira SD (2023) Fluoroquinolone and beta-lactam antimicrobials induce different transcriptome profiles in Salmonella enterica persister cells. Sci Rep 13(1):18696

52. Maragkoudakis PA, Zoumpopoulou G, Miaris C, Kalantzopoulos G, Pot B, Tsakalidou E (2006) Probiotic potential of Lactobacillus strains isolated from dairy products. Int Dairy J 16(3):189–199

53. Siepmann FB, De Almeida BS, Gomes TA, Waszczynskyj N, Spier MR (2022) Biochemical, microbiological and technological characteristics of type II sourdoughs produced with lactic acid bacteria single-strains. Cienc Agrar 43:693–712

# Chapter 6

# Preparation of Suspensions of Bacterial Spores

## Leonardo do Prado-Silva and Dionisio Pedro Amorim Neto

### Abstract

Bacterial spores are ubiquitous in the environment and can contaminate food through various means. Notably, spore-forming bacteria, like *Bacillus* and *Clostridium* species, have complex structures that confer high resistance to heat and chemicals. Spore production plays a crucial role in microbiology, offering diverse applications in food science and beyond due to spores' resistance to adverse conditions. With the knowledge of bacterial spore cultivation, food scientists can leverage spores' unique properties to explore a series of potential applications. The present protocol provides insights into the fundamental steps and considerations involved in producing bacterial spore suspensions.

**Key words** Bacterial spores, Food contamination, Spore production, Resistance, Spoilage

## 1 Introduction

Bacterial spores are widely distributed in the environment and can contaminate food through direct contact with soil and/or water. Bacterial spores are highly resistant to heat and chemicals due to their complex structure, which includes reduced water content, DNA saturation with small acid-soluble spore proteins (SASPs), and an inner membrane that protects the spore core [1]. Among the spore-forming bacteria, species belonging to the genera *Bacillus* and *Clostridium* are particularly notable, including *B. subtilis* [2], *B. cereus* [3], *B. licheniformis* [4], *C. botulinum* [5], *C. perfringens* [3], *C. sporogenes* [6], and *C. difficile* [7]. Bacterial spores can substantially affect food safety and spoilage. Therefore, spore production has become an important technique in microbiology, offering a wide range of applications in food science and other fields due to their durability and resistance to adverse environmental conditions. This process involves the cultivation, collection, and preservation of bacterial spores, which are resilient, dormant forms of bacteria. By understanding the methods and principles behind bacterial spore production, food scientists can harness their unique properties to evaluate the resistance of thermal

Anderson S. Sant'Ana (ed.), *Heat Resistant Microorganisms in Food Safety and Spoilage*, Methods and Protocols in Food Science, https://doi.org/10.1007/978-1-0716-5280-0_6,

processing of food [8] and innovative food processing technologies [9], shelf life [10], probiotic properties [11], and antimicrobial therapies [12]. This protocol explores the fundamental steps and considerations in the production of a bacterial spore suspension, highlighting its significance and diverse applications.

## 2 Materials

### 2.1 Media

1. *Nutrient Broth (NB)*. In g/L: 5.0 g of peptone, 5.0 g of sodium chloride, 1.5 g of beef extract, 1.5 g of yeast extract, and 1 L of distilled water. Under stirring, adjust the final pH to 5.4 ± 0.2 using 1 N HCl. Sterilize by autoclaving at 15 lb pressure (121 °C) for 15 min. If the solid version of the medium (Nutrient Agar, NA) is needed, just add 15 g of agar per liter. Wait for the melted media to reach an approximate temperature of 55 °C before pouring it into sterile Petri (90 × 15 mm diameter). Store at 4 °C for up to 30 days.

### 2.2 Reagents and Solutions

1. *Deionized water*. Sterilize by autoclaving at 15 lb pressure (121 °C) for 15 min. Store at 4 °C until use.
2. To prepare 1 L of *peptone water 0.1%(wv)*, dissolve 1.0 g peptone in 1000 mL of distilled water. Sterilize by autoclaving at 15 lb pressure (121 °C) for 15 min. Store at 4 °C until use.

## 3 Methods

### 3.1 Reactivation of a Cryopreserved Culture

1. Add 15 μL of the cryopreserved culture to a tube containing 5 mL of NB broth (*see* **Note 1**).
2. Vortex for 10 s to mix, then incubate at 30 °C for 24 h (*see* **Note 2**).
3. Streak the culture onto a NA agar plate to obtain isolated colonies. Incubate the plate at 30 °C for 24–48 h (*see* **Note 1**).

### 3.2 Propagation to Stabilize the Basal Phenotype

1. Transfer three to four colonies to 10 mL of NB, vortex for 10 s, and incubate at 30 °C for 24 h.
2. Add 100 μL of the preculture to 10 mL of NB, vortex for 10 s, and incubate at 30 °C for 24 h.
3. Repeat Step 2 once more.

### 3.3 Increase in Biomass Production and Inducing Sporulation

1. Finally, add 100 μL of the preculture to Petri dishes containing sporulation-promoting agar, such as NA supplemented with 3.6 μg/L of manganese chloride tetrahydrate (*see* **Note 3**), and spread evenly across the surface using a sterile Drigalski loop. Incubate at 30 °C for 5–10 days (*see* **Note 4**).

2. Periodically (every 3 days), check the plates for contamination.
3. At the same intervals, also check the sporulation status. To do this, use a sterile inoculation loop and transfer a small portion of the biomass to a 1.5 mL sterile microtube containing 1 mL of deionized water. Homogenize the suspension with a vortex for 15 s and view on a glass coverslip in a phase-contrast microscope (*see* **Note 5**).
4. Repeat Steps 2 and 3 successively until achieving a culture with a bacterial sporulation rate greater than 95%.

### 3.4 Spore Collection

1. Collect the total biomass (including spores and remaining vegetative cells) by washing the plate surface with 10 mL portions of sterile deionized water. Gently scrape the surface using a sterile Drigalski loop (*see* **Note 6**). Transfer the resulting suspension to a centrifuge tube.
2. Repeat Step 1 three more times (or until all the biomass has been collected).
3. Centrifuge the suspension under 8000 × g for 10 min at 4 °C (*see* **Note 7**).

### 3.5 Checking and Adjusting the Spore Concentration

1. Resuspend the pellet in 10 mL of deionized water (alternatively, vegetative cell inactivation methods can be performed (*see* **Note 8**) or adjust the volume as needed to achieve the desired final spore concentration. Determine the spore count after serial dilution and plating on NA, followed by incubation at 30°C for 24h. Dispense the suspension into cryotubes, with each tube containing 1 mL. Store the cryotubes at −80 °C for an undetermined period. It is recommended that the viability of spores be assessed by plate count every 6 months.
2. Before using, confirm the spore concentration using a culture-dependent method. Start by subjecting the suspension to heat treatment at 80 °C for 10 min in a water bath to activate spore germination (*see* **Note 9**). Immediately cool the cryotube by placing it in an ice bath. Prepare serial dilutions using 0.1% (w/v) peptone water and plate the diluted suspensions using the spread plate method. Incubate the plates at 30 °C for 24 h. and count the number of colonies. Express the result as log spores per milliliter.

## 4 Notes

1. It is important to note that the corresponding broth and agar should be used, such as nutrient broth and then nutrient agar, to avoid variations in metabolism during the passages.

2. Basal incubation times and temperatures can vary according to the microorganism and the presentation of the medium (whether broth or agar). Some conditions for *Bacillus* spp. (37 °C/15 h, in broth [13]), for *B. cereus* (32 °C/24 h [14]), for *B. subtilis* (37 °C for 48 h [15]), and for *A. acidoterrestris* (45 °C / overnight [16], or 24 h [17]) have already been reported.
3. Bacteria sporulation media includes mostly Nutrient Agar supplemented with 3.6 μg/L manganese chloride tetrahydrate [13] or manganese sulfate (10 ppm) [9, 18, 19]. However, other formulations have also been described, such as Nutrient broth: 16 g/L, Agar: 17 g/L, KCl: 2 g/L, MgSO4: 0.5 g/L, Ca(NO3)2: 0.164 g/L, MnCl2·4H2O: 0.0198 g/L, FeSO4: 0.00015 g/L, Glucose: 10 g/L [15] or Nutrient broth: 13 g/L, MgSO4.7H2O: 0.51 g/L, KCl: 0.97 g/L, CaCl2.2H2O: 0.2 g/L, MnSO4.H2O: 0.003 g/L, FeSO4.7H2O: 0.00055 g/L, Agar: 15 g/L, MnSO4.H2O: 3 mg/L, FeSO4.7H2O: 0.55 mg/L [20, 21].
4. The sporulation period can also vary. In this sense, it has already been described for Bacilli spp. 37 °C for 10 days [13], for *B. cereus* 32 °C for 5 days [14], 30 °C for 5–10 days [21], or 30 °C for 30 days [9, 18], for *B. subtilis* A163 or for *B. subtilis* AdHC1, *B. licheniformis* VBBBa 03–25 and AS 1·813, and *B. coagulans* ATCC 10545, 30 °C and 37 °C, respectively, for up to 5–10 days [15, 20], and for *A. acidoterrestris* 45 °C for 30 days for [17].
5. The differentiation of spores and vegetative cells can also be done using the Schaeffer–Fulton method and visualized under light microscopy (see [22] for more details).
6. Alternatively, a glass slide can be used to scrape the biomass [13].
7. Other centrifugation conditions can also be used, such as 5,900 × g for 15 min at 4 °C [14], 1500 × g for 20 min at 4 °C [9, 17, 18], 4000 × g 10 min at 23 °C [23], or 21,000 × g for 15 min at 4 °C [16].
8. Since spore collection is performed when the culture reaches >95% sporulation rate, some studies do not carry out spore purification steps [20, 21, 23]. In a simple way, however, others apply heat treatments ranging from 80 °C for 10 min [17] and 80 °C for 30 min [9] to eliminate vegetative cells. More elaborate approaches, however, include resuspending the pellet in 70% (v/v) ethanol, followed by heat treatment at 80 °C for 20 min and then an ice bath [14], centrifugation cycles using 50% (v/v) ethanol [16] or physical and enzymatic cycles simultaneously (see [15] for more details).

9. Although heat treatment at 80 °C for 10 min is widely used to induce bacterial spore germination [16, 23, 24], other conditions, like 75 °C for 20 min [18, 21], have also been reported.

## References


1. Setlow P, Christie G (2023) New thoughts on an old topic: secrets of bacterial spore resistance slowly being revealed. Microbiol Mol Biol Rev 87(2). https://doi.org/10.1128/mmbr.00080-22
2. Setlow P (2006) Spores of *Bacillus subtilis*: their resistance to and killing by radiation, heat and chemicals. J Appl Microbiol 101:514–525. https://doi.org/10.1111/j.1365-2672.2005.02736.x
3. Newell K, Helfrich K, Isernhagen H et al (2024) Multipathogen outbreak of *Bacillus cereus* and *Clostridium perfringens* among Hospital Workers in Alaska, August 2021. Public Health Rep 139:195–200. https://doi.org/10.1177/00333549231170220
4. Baril E, Coroller L, Couvert O et al (2012) Modeling heat resistance of *Bacillus weihenstephanensis* and *Bacillus licheniformis* spores as function of sporulation temperature and pH. Food Microbiol 30:29–36. https://doi.org/10.1016/j.fm.2011.09.017
5. Mertaoja A, Mascher G, Nowakowska MB et al (2023) Cellular and population strategies underpinning neurotoxin production and sporulation in *Clostridium botulinum* type E cultures. MBio 14. https://doi.org/10.1128/MBIO.01866-23/SUPPL_FILE/MBIO.01866-23-S0001.DOCX
6. Naim F, Zareifard MR, Zhu S et al (2008) Combined effects of heat, nisin and acidification on the inactivation of *Clostridium sporogenes* spores in carrot-alginate particles: from kinetics to process validation. Food Microbiol 25:936–941. https://doi.org/10.1016/j.fm.2008.06.005
7. Serrano M, Martins D, Henriques AO (2024) *Clostridioides difficile* sporulation. Adv Exp Med Biol 1435:273–314
8. Oliveira RBA, Baptista RC, Chincha AAIA et al (2018) Thermal inactivation kinetics of *Paenibacillus sanguinis* 2301083PRC and *Clostridium sporogenes* JCM1416MGA in full and low fat "requeijão cremoso". Food Control 84:395–402. https://doi.org/10.1016/j.foodcont.2017.08.030
9. do Prado-Silva L, Alvarenga VO, Braga GÚL, Sant'Ana AS (2021) Inactivation kinetics of *Bacillus cereus* vegetative cells and spores from different sources by antimicrobial photodynamic treatment (aPDT). LWT 142:111037. https://doi.org/10.1016/J.LWT.2021.111037
10. André S, Vallaeys T, Planchon S (2017) Sporeforming bacteria responsible for food spoilage. Res Microbiol 168:379–387. https://doi.org/10.1016/j.resmic.2016.10.003
11. Soares MB, Almada CN, Pereira EPR et al (2023) Review – Sporeforming probiotic bacteria: characteristics, health benefits, and technological aspects for their applications in foods and beverages. Trends Food Sci Technol 138:453–469. https://doi.org/10.1016/j.tifs.2023.06.029
12. Anumudu C, Hart A, Miri T, Onyeaka H (2021) Recent advances in the application of the antimicrobial peptide Nisin in the inactivation of Sporeforming bacteria in foods. Molecules 26:5552. https://doi.org/10.3390/molecules26185552
13. Furukawa S, Narisawa N, Watanabe T et al (2005) Formation of the spore clumps during heat treatment increases the heat resistance of bacterial spores. Int J Food Microbiol 102:107–111. https://doi.org/10.1016/j.ijfoodmicro.2004.12.004
14. Mohammadi B, Pérez Reyes ME, Smith SA (2024) Survival, growth, and toxin production of *Bacillus cereus* during cooking and storage of fresh Rice noodles. J Food Prot 87:100239. https://doi.org/10.1016/J.JFP.2024.100239
15. Vincent LN, Fayolle EC, Hodyss R et al (2024) Bacterial spore morphology remains highly recognizable after exposure to simulated Enceladus and Europa surface conditions. Commun Earth Environ 5(1):1–10. https://doi.org/10.1038/s43247-024-01872-z
16. Silva FVM, Gibbs P (2001) *Alicyclobacillus acidoterrestris* spores in fruit products and design of pasteurization processes. Trends Food Sci Technol 12:68–74. https://doi.org/10.1016/S0924-2244(01)00070-X
17. do Prado-Silva L, Brancini GTP, Bachmann L et al (2025) Controlling *Alicyclobacillus acidoterrestris* spores contamination on orange surfaces using different photosensitizers and light sources. Food Control 167:110798. https://doi.org/10.1016/J.FOODCONT.2024.110798

18. Alvarenga VO, Campagnollo FB, AKR P et al (2018) Quantifying the responses of three *Bacillus cereus* strains in isothermal conditions and during spray drying of different carrier agents. Front Microbiol 9:381252. https://doi.org/10.3389/fmicb.2018.01113
19. Pflug IJ (1999) Microbiology and engineering of sterilization processes, 10th edn. University of Minnesota, Minneapolis
20. Cazemier AE, Wagenaars SFM, Ter Steeg PF (2001) Effect of sporulation and recovery medium on the heat resistance and amount of injury of spores from spoilage bacilli. J Appl Microbiol 90:761–770. https://doi.org/10.1046/J.1365-2672.2001.01302.X
21. Faille C, Lebret V, Gavini F, Maingonnat JF (1997) Injury and lethality of heat treatment of *Bacillus cereus* spores suspended in buffer and in poultry meat. J Food Prot 60:544–547. https://doi.org/10.4315/0362-028X-60.5.544
22. da SN, Taniwaki MH, Junqueira VCA et al (2019) Basic techniques for the detection of the presence/absence of microorganisms. In: Microbiological examination methods of food and water, pp 57–64. https://doi.org/10.1201/9781315165011
23. Friedrich LM, Goodrich-Schneider R, Parish ME, Danyluk MD (2009) Mitigation of *Alicyclobacillus* spp. spores on food contact surfaces with aqueous chlorine dioxide and hypochlorite. Food Microbiol 26:936–941. https://doi.org/10.1016/J.FM.2009.06.011
24. Dos Anjos MM, Ruiz SP, Nakamura CV, De Abreu Filho BA (2013) Resistance of *Alicyclobacillus acidoterrestris* spores and biofilm to industrial sanitizers. J Food Prot 76:1408–1413. https://doi.org/10.4315/0362-028X.JFP-13-020

# Chapter 7

# Preparation of Suspensions of Filamentous Fungi and Yeasts

**Dionisio Pedro Amorim Neto and Luísa Freire**

## Abstract

The preparation of microbial suspensions is the initial step for conducting experiments to evaluate the inactivation of fungi and yeasts, their potential for growth in food products, and the production of metabolites by these microorganisms. Accurate microbial suspension preparation is essential to ensure reliable results in food microbiology research. This chapter provides detailed instructions on preparing microbial suspensions, covering both vegetative cells and fungal spores from molds and yeasts. Various experimental methods and conditions are discussed to accommodate diverse investigational needs.

**Key words** Inoculum, Vegetative cell, Ascospore, Spore

## 1 Introduction

Molds and yeasts commonly contaminate foods like juices, bread, dairy products, and fruits, causing spoilage [1, 2]. Understanding how these fungi tolerate stressors such as heat, pH, and water activity is crucial for extending the shelf life of various food products, predicting the production of toxic metabolites, and assessing their safety [3, 4]. Additionally, identifying and isolating species of interest allow these microorganisms to be used in other studies, as well as enabling biological control and future applications in biotechnology. Growth models can be applied to evaluate fungal growth kinetics and extrapolate their behavior and physiology in different foods, aiming to enhance food quality and safety. However, it is essential to consider the specific characteristics of each microorganism, particularly their biological and strain variability [5]. In this sense, accurate preparation and consistent composition of microbial suspensions are essential to ensure reliable results and precise interpretations in food microbiology research.

This chapter outlines the procedures for preparing mold and yeast suspensions. These suspensions are indispensable for a wide

Anderson S. Sant'Ana (ed.), *Heat Resistant Microorganisms in Food Safety and Spoilage*, Methods and Protocols in Food Science, https://doi.org/10.1007/978-1-0716-5280-0_7,

range of experiments, including investigations into the microorganisms' heat resistance and their ability to grow in food products over a specific storage condition (challenge tests), among others.

Generally, preparing a microbial suspension involves several steps: (i) reactivation of a cryopreserved culture, (ii) propagation to stabilize the basal phenotype, (iii) increase in biomass production, (iv) bioproduct collection, and finally (v) checking and adjusting the cellular and/or spore concentration. Alternatively, this preparation flow can be adapted to include producing and collecting spores generated by molds and yeasts.

Essentially, spores in molds and yeasts result from a reproductive process. Environmental factors such as nutrient scarcity or abiotic stressors, like temperature, humidity, etc. [6, 7], can stimulate this reproductive process in fungi. In yeasts, nonfermentable carbon, like acetate, can also induce sporulation, which occurs through meiosis and the formation of ascospores from the cytoplasm of a mother cell [8]. In molds, ascospores are sexual spores of *Ascomycetes*, formed in specialized structures, the ascomata [9, 10]. When ascospores survive heat treatments and find suitable conditions, they can germinate and grow, resulting in the visible formation of mycelium. In addition, some species may be potentially toxigenic [11]. Finally, spore formation gives fungi a significant adaptive advantage, enabling them to disperse widely and withstand adverse conditions, causing food spoilage and contamination [12, 13].

## 2 Materials

### 2.1 Media

1. *Malt Extract Agar (MEA)*. In g/L: 20.0 g of malt extract, 1.0 g of casein peptone, 30.0 g of glucose, 20 g of agar, and 1 L of distilled water. Sterilize by autoclaving at 15 lb pressure (121 °C) for 15 min. Wait for the melted media to reach an approximate temperature of 55 °C before pouring it into sterile Petri (90 × 15 mm diameter). Store at 4 °C for up to 30 days.
2. *Yeast, Peptone, and Dextrose (YPD) broth*. In g/L: 20 g of peptic digest of animal tissue, 10 g of yeast extract, 20 g of dextrose, and 1 L of distilled water. Under stirring, adjust the final pH to 6.5 ± 0.2 using 1 N HCl. Divide the medium into flasks containing 5, 10, and 100 mL of broth and sterilize by autoclaving at 15 lb pressure (121 °C) for 15 min. Store at 4 °C for up to 30 days.
3. *Yeast, Peptone, and Dextrose (YPD) agar*. In g/L: 20 g of peptic digest of animal tissue, 10 g of yeast extract, 20 g of dextrose, 15 g of bacteriological agar, and 1 L of distilled water. Under stirring, adjust the final pH to 6.5 ± 0.2 using 1 N HCl. Sterilize by autoclaving at 15 lb pressure (121 °C) for 15 min.

Wait for the melted media to reach an approximate temperature of 55 °C before pouring it into a sterile Petri (90 × 15 mm diameter). Store at 4 °C for up to 30 days.

4. *Ascospore agar.* In g/L: 2.5 g of yeast extract, 1 g of dextrose, 10 g of potassium acetate, 30 g of bacteriological agar, and 1 L of distilled water. Under stirring, adjust the final pH to 6.4 ± 0.2 using 1 N HCl. Sterilize by autoclaving at 15 lb pressure (121 °C) for 15 min. Wait for the melted media to reach an approximate temperature of 55 °C before pouring it into a sterile Petri (90 × 15 mm diameter). Store at 4 °C until use.

### 2.2 Reagents and Solutions

1. *Distilled water.* Sterilize by autoclaving at 15 lb pressure (121 °C) for 15 min. Store at 4 °C until use.
2. To prepare 1 L of *0.1% Tween 80*, dissolve 1.5 mL of Tween 80 in 998.75 mL of distilled water. Sterilize by autoclaving at 15 lb pressure (121 °C) for 15 min. Store at 4 °C until use.
3. To prepare 1 L of *1X PBS* (0.1 M PBS), dissolve 7.65 g of sodium chloride (NaCl), 0.724 g of disodium hydrogen phosphate anhydrous ($Na_2HPO_4$), and 0.210 g of potassium dihydrogen phosphate ($KH_2PO_4$) in 900 mL of distilled water. Adjust the final pH to 7.4 ± 0.2 using 1 N NaOH and top up to 1 L. Divide the solution into 10 mL bottles to adjust the suspension and the rest in 500 mL bottles. Sterilize by autoclaving at 15 lb pressure (121 °C) for 15 min. Store at 4 °C until use.
4. *Methylene blue solution.* To prepare a 0.1% (w/v) methylene blue solution, weigh out 0.1 g of the dye and slowly add 100 mL of distilled water while stirring. Continue stirring until the solution is homogeneous and the powder is completely dissolved. Then, filter the solution using a 0.22 μm membrane. It can be stored at room temperature until needed.

## 3 Methods

### 3.1 Filamentous Fungi Conidial Suspension

1. If the isolate is preserved in an inclined tube, using a sterile instrument (inoculation loop or wooden stick), remove a portion of conidia from the surface of the mycelium and transfer it directly to a point in the center of a sterile Petri containing solidified MEA culture medium (*see* **Note 1**).
2. If the isolate is preserved on a filter paper (usually in the form of 0.5 cm diameter discs or 0.25 cm$^2$ strips) impregnated with conidia and frozen in microcentrifuge tubes, using sterile forceps, remove a fragment of the filter paper containing the

conidia and transfer it directly to the center of a sterile Petri containing solidified MEA culture medium.

3. Incubate the plate at 25 °C for 5–7 days (*see* **Note 2**).
4. Add 5 mL of 0.1% Tween 80 solution directly onto the sterile Petri containing the isolate. Using a sterile Drigalsky loop, scrape the conidia from the surface of the culture medium, gently mixing them with the solution (*see* **Note 3**).
5. Filter the suspension obtained using a sterile gauze and transfer the filtrate to a 50 mL Falcon tube.
6. Repeat this procedure two more times to ensure complete collection of conidia.
7. Centrifuge the suspension at 11,962.6 × g for 15 min at 5 °C (*see* **Note 4**).
8. After centrifugation, carefully discard the supernatant, preserving the pellet containing the conidia.
9. Add 5 mL of sterile distilled water to the tube containing the pellet and gently resuspend. Repeat the centrifugation and washing process two more times.
10. After the last wash, carefully discard the supernatant, preserving the pellet containing the conidia, and adjust to the desired concentration with sterile water. Homogenize to obtain the final conidial suspension (*see* **Note 5**).
11. To determine the final concentration of conidia in the fungal suspension, count the conidia in a cell-counting hemocytometer (Neubauer chamber). Standardized suspensions can be stored at −18 °C for up to 6 months (*see* **Notes 6** and **7**).

### 3.2 Heat-resistant Filamentous Fungi Spore Suspension

1. Please see item 3.1, Methods, steps 1 to 2 (*see* **Note 8**).
2. Incubate the plate at 30 °C for 30 days to ensure sporulation (*see* **Note 9**).
3. Add 8 mL of 0.1% Tween 80 to each plate to detach the spores (*see* **Note 10**).
4. Scrape the surface of the medium with a sterile Drigalsky loop to release the spores.
5. Filter the suspension through three layers of sterile gauze to remove hyphae (*see* **Note 4**).
6. Centrifuge the filtrate at 8000 × g for 15 min at 4 °C (*see* **Note 11**).
7. Resuspend the pellet in 10 mL of sterile distilled water.
8. Sonicate the suspension (30-s intervals) for up to 4 min, maintaining the temperature between 0 °C and 4 °C to separate the ascospore clumps.

9. For storage, transfer the final suspensions to glass vials containing sterile glass beads and store at 2 °C for 6 months.
10. To determine the spore concentration, a thermal shock must be performed prior to inoculation: Transfer 1 mL of the prepared suspension to a heat-resistant glass tube. Place the tube in a preheated water bath at 80 °C for 10 min to activate the spores. Immediately after heating, cool the tubes in an ice bath to stop the activation (*see* **Note 12**).
11. Determine the cell counts of filamentous fungi by direct methods such as serial dilution and surface spreading on MEA culture medium after incubation of the plates for 5–7 days at 30 °C (*see* **Note 13**).

### 3.3 Yeast Suspension

1. Transfer an aliquot (or a 10 μL loop) of the cryopreserved culture to a tube containing 5 mL of YPD broth (*see* **Note 14**).
2. Vortex for 10 s to mix; then, incubate at 30 °C for 24–48 h (*see* **Note 15**).
3. Streak the culture onto a YPD agar plate to obtain isolated colonies. Incubate the plate at 30 °C for 24–48 h (*see* **Note 16**).
4. Transfer one to three colonies to 10 mL YPD, vortex for 10 s, and incubate at 30 °C for 24 h.
5. Add 100 μL of preculture to 10 mL of YPD, vortex for 10 s, and incubate at 30 °C for 24 h.
6. Repeat step 5 once more.
7. Finally, add 100 μL of the preculture to a flask containing 100 mL of YPD broth and incubate with shaking at 30 °C, 200 RPM for 24–48 h (*see* **Note 17**).
8. Centrifuge the cell suspension at 2000 × g for 10 min at 4 °C to pellet the yeast cells. After centrifugation, discard the supernatant (*see* **Note 18**).
9. Resuspend the pellet in an appropriate volume of 1X PBS to obtain a height of approximately 1 cm above the pellet. Homogenize the suspension by vortexing at maximum speed. Repeat the centrifugation under the same conditions (*see* **Notes 19** and **20**).
10. Repeat step 9 twice more.
11. Resuspend the pellet in an appropriate volume of 1X PBS to obtain a height of approximately 1 cm above the pellet. Homogenize the suspension by vortexing at maximum speed and reserve.
12. Add aliquots of the yeast inoculum to a tube containing 10 mL of 1X PBS. Vortex for 10 s after each addition until the desired cell concentration is reached (*see* **Note 21**).

### 3.4 Yeast Ascospore Suspension

1. Please *see* Subheading 3.3, Methods, steps 1 to 3.
2. Please *see* Subheading 3.3, Methods, steps 4 to 6.
3. Using a sterile swab, streak a plate of Ascospore Agar, covering the entire surface. Incubate at 25 °C for 9–10 days (*see* **Notes 22** and **23**).
4. Periodically (every 3 days), check the plates for potential contaminations.
5. At the same intervals, also check the progress of sporulation. To do this, use a sterile inoculation loop and transfer a small portion of the biomass to a 1.5 mL sterile microtube containing 1 mL of methylene blue solution. Homogenize the suspension with a micropipette (doing up-and-down movements) and visualize it under a microscope using a hemocytometer (*see* Subheading 3.3, **Note 21**).
6. Collect all the biomass using a sterile Drigalski loop (or a cell scraper) and transfer it to a 50 mL sterile Falcon tube.
7. Top the tube with sterile distilled water (*see* **Note 24**), leaving approximately 1–2 cm of headspace above the biomass. Vortex for 5 s and centrifuge at 3000 × g for 5 min at 4 °C (*see* **Note 25**).
8. Repeat step 7 one more time.
9. Resuspend the ascospore pellet in 10 mL of sterile distilled water and vortex until fully dispersed.
10. Add aliquots of ascospore inoculum to a new tube containing 10 mL of sterile distilled water. Vortex for 10 s after each addition until the desired cell concentration is reached.
11. To count the ascospores, in a 1.5 mL sterile microtube pipette, add 10 μL of the ascospore suspension into 90 μL of methylene blue solution. Mix the solution by pipetting up and down, then count the ascospores using a hemocytometer (*see* Subheading 3.3, **Note 21**).
12. Store the ascospore suspension at 4 °C for no more than 1 week.

## 4 Notes

1. *MEA* culture medium is commonly used for growing fungi, especially from the genera *Aspergillus* and *Penicillium* [5, 9, 14, 15]. However, other media can be used for this purpose, especially for other genera such as *Fusarium*, with emphasis on *Potato Dextrose Agar* (PDA) [16] in g/L: 4.0 g of Potato Extract—equivalent to 200.0 g of infusion from potatoes, 20.0 g of dextrose, 15.0 g of agar.

2. In some cases, prolonged incubation is necessary for adequate growth of the isolate. Different genera may require different ideal incubation times and temperatures. Other conditions include: Room temperature (22–25 °C) for 10 days in PDA culture medium for *F. oxysporum*, *F. proliferatum*, and *F. verticillioides* [17]; 35 °C for 3 to 5 days in PDA culture medium for *Aspergillus fumigatus, Aspergillus terreus, Aspergillus flavus, Fusarium solani, Paecilomyces variotii, Fusarium verticillioides, Aspergillus sydowii, Paecilomyces lilacinus, Fusarium oxysporum, Aspergillus nidulans, Trichoderma viride, Aspergillus versicolor, Acremonium kiliense, Cladophialophora* spp., *Aspergillus ochraceus* [18]; for *Simplicillium lamellicola*, incubation at room temperature (22–25 °C) for 3 weeks on PDA medium, followed by obtaining a hyphal suspension [19]. The inoculum is then transferred to a new PDA plate and incubated under mixed fluorescent and ultraviolet (UV) lighting on a 12 h light and 12 h dark cycle for 1 week.
3. Typically, for hydrophobic fungi, such as *Aspergillus* sp., *Paecilomyces* sp., *Penicillium* sp., *Scopulariopsis* sp., and *Trichoderma* sp., a 0.1% Tween 80 solution is used to facilitate dispersion. For hydrophilic genera, such as *Fusarium* and *Scedosporium*, only sterile distilled water is used, ensuring better adaptation to the hydrophilic characteristics of these genera [18].
4. Centrifugation parameters can be adjusted for different cell types and volumes, improving the separation efficiency, such as 11,000 × g for 15 min at 4 °C [20] or 8725 × g for 15 min [21].
5. The amount of water added depends on the desired concentration and the number of conidia obtained in the pellet. For example, for *Aspergillus niger* grown in MEA culture medium for 7 days at 25 °C, the addition of 40 mL of water allows obtaining a suspension with a concentration of $10^6$ conidia/mL [5]. To check the concentration, count the conidia in a cell-counting hemocytometer (Neubauer chamber).
6. Stored spore suspensions should be periodically checked by microscopy [22] and enumeration on MEA or PDA to ensure that their concentration remains stable.
7. Cell counts of filamentous fungi can be determined by direct methods such as serial dilution and surface spreading on solid media. Plates should be observed daily for the presence of growth, and colonies should be counted as soon as possible after visible growth is observed. Indirect methods can also be used, such as measuring optical density at 530 nm using a spectrophotometer [17, 18]. It is important to note that the size and color of conidia can vary between species. Therefore, it

is recommended to construct a standard curve correlating absorbance with the cell-counting hemocytometer and direct cell counts to obtain a specific standard for each suspension.

8. To increase the number of ascospores produced by osmophilic species such as *E. repens* [11], osmophilic media can also be used, such as M40Y (In g/L: 400 g of sucrose, 20 g of malt extract, 5 g of yeast extract, and 1 L of distilled water). PDA was also used for reactivation of cultures of *Aspergillus hiratsukae*; *Aspergillus thermomutatus*, *Chaetomium globosum*; *Chaetomium flavoviride*, *Talaromyces bacillisporus*, and *Talaromyces trachyspermus* [23].
9. In some cases, other incubation conditions are necessary for adequate growth of the isolate; for example, *E. repens* [11] was grown at 24 °C for 4 weeks in a medium recommended for osmophilic fungi. For *Talaromyces avellaneus* [24], *Aspergillus hiratsukae*, *Aspergillus thermomutatus*, *Chaetomium globosum*, *Chaetomium flavoviride*, *Talaromyces bacillisporus*, and *Talaromyces trachyspermus* [23], the incubation time was 8 weeks. Sant'Ana et al. (2009) suggest that culture reactivation be performed in Roux bottles containing 200 mL of sterile formulated MEA and incubated for 30 days at 30 °C.
10. Instead of 0.1% Tween 80, isotonic NaCl solution with 0.1% Tween 80 can be used to prevent clumping of spores [11]. When growth is performed in Roux bottles, it is suggested to use 20 mL sterile distilled water to collect the ascospores [25].
11. Before centrifugation, shaking the suspension with glass beads (2 mm in diameter) for 3 min can be used as an alternative to break the asci, according to Michener and King (1974) [26]. Racchi et al. (2020) [23] suggest shaking the suspension with glass beads for 5 min. Voldrich et al. (2004) [24] suggest shaking with glass beads in an ultrasonic bath and filtering through sterile glass wool. Maneeboon et al. (2023) [27] suggest first centrifuging at 4000 × g for 10 min, discarding the supernatant and washing three times with sterile 0.1% (w/v) peptone and then adding sterile glass beads to the ascospore suspension and mixing for 2 min using a vortex mixer to break up ascospore clusters and then centrifuging again at 4000 × g for 10 min. For the separation of ascospores from vegetative cells, centrifugation in 34% CsCl solution for 20 min at 1100 × g can be used [24]. Another condition may be the use of a 34% CsCl solution (containing 0.1% Tween 80) and centrifugation for 5 min at 2600 × g [11]. After centrifugation, the upper part of the solution containing the conidiospores is removed. This procedure should be repeated at least three times until no more conidiospores remain. In the second

purification step, the ascospores should be layered on a 38% CsCl solution (containing 0.1% Tween 80) and centrifuged for 20 min at 1100 × g. Viable ascospores were concentrated in a thin band at the top of the centrifuge tube, while nonviable spores were deposited at the bottom of the tube. The band containing the ascospores should be removed and washed three times to remove all CsCl. When growth is performed in Roux bottles, the following centrifugation conditions were used: 11962.6 × g and 4 °C for 15 min [25].

12. Other conditions for thermal shock can be used, for example, in isotonic NaCl (containing 0.1% Tween 80), at 60 °C for 15 min to *E. repens* [11], and 65 °C for 15 min in a thermostatic bath to eliminate fungal conidia and vegetative cells of *Aspergillus laciniosus*, *A. chevalieri*, *A. denticulatus*, *A. siamensis*, *Hamigera pallida*, and *Talaromyces macrosporus* [27]. Heat shocks at 75 °C for 5 min were used for *B. fulva* and 20 min for *B. nivea* [25].
13. Osmophilic media can also be used, such as M40Y [11]. PDA containing 50 mg/L of rose bengal and 100 mg/L of chloramphenicol was used to enumerate the species: *Aspergillus laciniosus*, *A. chevalieri*, *A. denticulatus*, *A. siamensis*, *Hamigera pallida*, and *Talaromyces macrosporus* [27].
14. *Yeast Extract Peptone Dextrose (YPD or YEPD) broth* proposed by Ausubel and collaborators (1994) [28] and Sherman (1991) [29] is commonly used in yeast cultivation [4, 30–35]. However, other nonselective media can be used for this purpose, such as the *Yeast Malt broth (YM broth, YMPG broth, or Wickerham Medium)* [6, 36–38] in g/L: 5 g of peptic digest of animal tissue, 3 g of yeast extract, 3 g of malt extract, and 10 g of dextrose, pH 6.2 ± 0.2 [39]. Sometimes, for *Wickerham Medium*, 1.5 g/L of $NaH_2PO_4$ could be added [3]. *Sabouraud Dextrose (SD) broth* [40] in g/L: 10 g of peptone and 20 g of dextrose, pH 5.6 ± 0.2 [41–43].
15. Different yeasts and/or strains may require different optimal incubation times and temperatures. Other conditions include, for example, 25 °C for 48 h to *Candida parapsilosis, Candida pseudointermedia, Candida tropicalis, Clavispora lusitanae, Cryptococcus albidus, Debaryomyces hansenii, Kluyveromyces marxianus, Meyerozyma guilliermondi, Pichia anomala, Pichia fermentans, Saccharomyces cerevisiae, Torulaspora delbrueckii, Trichosporon cutaneum* [30], and *S. cerevisiae* [3]; 27 °C for 24–48 h to *Saccharomyces cerevisiae* [40]; 28 °C for 48 h to *Meyerozyma guilliermondii, Debaryomyces hansenii, Wickerhamomyces anomalus, Candida zemplinina, Torulaspora delbrueckii, Candida orthopsilosis, Meyerozyma caribbica, Candida glabrata, Candida tropicalis*, and *Zygosaccharomyces rouxii* [4].

16. Using agars with the same composition as the broths (used in the previous steps) is crucial to avoid microbial stress due to abrupt changes in the nutritional conditions of the culture medium and ensure consistent results. This applies to all stages of microbial cultivation during the suspension preparation. In this sense, 15–20 g of bacteriological agar per liter can be added.
17. This variable can be adjusted once the time needed to reach the desired cell concentration is known.
18. Centrifugation parameters can be adjusted for different cell types and volumes, improving separation efficiency, such as 2400 × g for 15 min [30] or 1475 × g, 5 min [40], for example.
19. Although it is recommended to wash the inoculum to remove the culture medium, some studies have directly added microbial suspension to the food (especially juices) without this step after adjusting the cell concentration [36].
20. Other diluents, such as 0.9 [34, 44] or even 0.85% (w/v) NaCl solution [4], or peptone 0.1% (w/v) solution [40] can be used for the collection of vegetative cells. However, the diluent choice should consider cell suspension applications. Buffers, like 1X PBS [45] or citrate–phosphate buffer [35], may be interesting to maintain and stabilize the pH of the suspension.
21. Yeast cell density can be determined by direct methods such as serial microdilution with and drop-plating on solid media (*see* "Drop plate technique" by da Silva and collaborators (2019) [46]) or hemocytometer counting (*see* the video protocol procedure by Jove (2023) [47]) and indirect methods, such as measuring optical density at 600 nm using a spectrophotometer. According to Treco and Winston (2001) [31], in wild yeasts, an absorbance of 0.1 at 600 nm corresponds to approximately $3 \times 10^6$ cells/mL, while an absorbance of 1 corresponds to about $3 \times 10^7$ cells/mL. It is important to note that cell size can vary between species and genotypes and in response to different stresses. Therefore, it is recommended to construct a standard curve correlating absorbance with direct cell counts to obtain a specific standard for each suspension.
22. Different solid media and incubation conditions can induce sporulation in various yeast species. Other options exist beyond the *Ascospore Agar* medium [48–50] for *Saccharomyces cerevisiae, Pichia farinosa, Kluyveromyces thermotolerans, Zygosaccharomyces bailii, Saccharomycodes ludwigii, Wickerhamomyces anomalus, Hansenula silvicola, Debaryomyces vanriji, Saccharomyces uvarum, Pichia manshurica, Wickerhamomyces anomalus,* and *Pichia subpelliculosa* used by Rachon and

collaborators (2021) [38]. Put and Jong (1982a,b) [51, 52] induced the production of ascospores in *S. bailii* spp., *S. cerevisiae*, *S. chevalieri*, and *S. uvarum* by cultivating them at 25 °C for 10–14 days in *2% Malt Extract Agar (MEA)* in g/L: 20 g of malt extract, 5 g of peptone, 15 g of bacteriological agar [53–55], pH 4.7; or *Kleyn's Acetate Agar* in g/L: 2.5 g of peptone, 0.6 g of sodium chloride, 5 g of sodium acetate trihydrate, 0.6 g of glucose, 20 g of bacteriological agar [56], pH 7.0. However, for other species like *K. bulgaricus and K. cicerisporus*, incubation could be carried out at 25 °C for 4–6 days in *Potato Dextrose Agar (PDA)* [51] in g/L: 4 g of potato infusion (infusion from 200 g potatoes), 20 g of dextrose, 15 g of bacteriological agar [41, 57], pH 5.5. Finally, Pomper, Daniels, and McKee (1954) promote the sporulation of *Saccharomyces cerevisiae* by inoculation in *Sporulation Media Agar (SPOM)* in g/L: 3.0 g of potassium acetate, 0.2 g of raffinose, 20 g of bacteriological agar, and 1 L of distilled water [58] and incubated it at 25 °C for 2–7 days [58].

23. Liquid media, like the *Sporulation Broth* in g/L: 10 g of potassium acetate, 1 g of yeast extract, 0.5 g of glucose, zinc sulfate 25 mg/L, used by Milani and collaborators (2015) [33], can also induce sporulation in yeasts like *S. cerevisiae* and *S. pastorianus* when incubated at 18 °C with 230 rpm shaking for 14 days. In another study, the sporulation of *Saccharomyces diastaticus* was achieved after incubation of 120 h at 25 °C and 200 rpm in the same Sporulation Broth without zinc sulfate [34].
24. While sterile distilled water [38, 59] is the most common diluent, 0.9% (w/v) NaCl solution [34, 51, 52] can also be used for washing and collecting ascospores.
25. Other centrifugation techniques, aside from those proposed by Rachon and collaborators (2021) [38], can be employed, such as 9700 × g for 1 min [33] or 13,000 × g for 10 min at 10 °C [51, 52], for example.

## References


1. de Carvalho TB, Silva BN, Tomé E, Teixeira P (2024) Preventing fungal spoilage from raw materials to final product: innovative preservation techniques for fruit fillings. Foods 13: 2669. https://doi.org/10.3390/FOODS13172669
2. Agirman B, Carsanba E, Settanni L, Erten H (2023) Exploring yeast-based microbial interactions: the next frontier in postharvest biocontrol. Yeast 40:457–475. https://doi.org/10.1002/YEA.3895
3. Laroche C, Fine F, Gervais P (2005) Water activity affects heat resistance of microorganisms in food powders. Int J Food Microbiol 97: 307–315. https://doi.org/10.1016/J.IJFOODMICRO.2004.04.023
4. Wang H, Hu Z, Long F et al (2016) The effects of stress factors on the growth of spoilage yeasts isolated from apple-related environments in apple juice. J Food Saf 36:162–171. https://doi.org/10.1111/JFS.12223
5. Freire L, Guerreiro TM, Pia AKR et al (2018) A quantitative study on growth variability and

production of ochratoxin A and its derivatives by A. carbonarius and A. niger in grape-based medium. Sci Rep 8:14573. https://doi.org/10.1038/s41598-018-32907-z

6. Raleigh C, Lawrence S, Rachon G (2024) Optimisation of Saccharomyces cerevisiae BRYC 501 ascospore formation and recovery for heat inactivation experiments. J Inst Brew 130:70–82. https://doi.org/10.58430/jib.v130i2.48
7. Manstretta V, Morcia C, Terzi V, Rossi V (2016) Germination of Fusarium graminearum ascospores and wheat infection are affected by dry periods and by temperature and humidity during dry periods. Phytopathology 106:262–269. https://doi.org/10.1094/PHYTO-05-15-0118-R/ASSET/IMAGES/LARGE/PHYTO-05-15-0118-R_F6-1457416829011.JPEG
8. Neiman AM (2005) Ascospore formation in the yeast Saccharomyces cerevisiae. Microbiol Mol Biol Rev 69:565–584. https://doi.org/10.1128/MMBR.69.4.565-584.2005/ASSET/186552EC-D66F-4EE7-AE47-4CBB30360BE7/ASSETS/GRAPHIC/ZMR0040520990008.JPEG
9. Pitt JI, Hocking AD (2009) Fungi and food spoilage. In: Fungi and food spoilage, pp 1–519. https://doi.org/10.1007/978-0-387-92207-2/COVER
10. Rico-Munoz E (2017) Heat resistant molds in foods and beverages: recent advances on assessment and prevention. Curr Opin Food Sci 17:75–83. https://doi.org/10.1016/J.COFS.2017.10.011
11. Eicher R, Ludwig H (2002) Influence of activation and germination on high pressure inactivation of ascospores of the mould Eurotium repens. Comp Biochem Physiol A Mol Integr Physiol 131:595–604. https://doi.org/10.1016/S1095-6433(01)00511-6
12. Dijksterhuis J (2019) Fungal spores: highly variable and stress-resistant vehicles for distribution and spoilage. Food Microbiol 81:2–11. https://doi.org/10.1016/J.FM.2018.11.006
13. Dijksterhuis J (2017) The fungal spore and food spoilage. Curr Opin Food Sci 17:68–74. https://doi.org/10.1016/J.COFS.2017.10.006
14. Garcia MV, da Pia AKR, Freire L et al (2019) Effect of temperature on inactivation kinetics of three strains of Penicillium paneum and P. roqueforti during bread baking. Food Control 96:456–462. https://doi.org/10.1016/J.FOODCONT.2018.10.002
15. Wigmann ÉF, Moreira RC, Alvarenga VO et al (2016) Survival of Penicillium spp. conidia during deep-frying and baking steps of frozen chicken nuggets processing. Food Microbiol 55:1–6. https://doi.org/10.1016/J.FM.2015.11.010
16. Li YC, Bi Y, Ge YH et al (2009) Antifungal activity of sodium silicate on Fusarium sulphureum and its effect on dry rot of potato tubers. J Food Sci 74:M213–M218. https://doi.org/10.1111/J.1750-3841.2009.01154.X
17. Caligiore-Gei PF, Valdez JG (2015) Adjustment of a rapid method for quantification of Fusarium spp. spore suspensions in plant pathology. Rev Argent Microbiol 47:152–154. https://doi.org/10.1016/J.RAM.2015.03.002
18. Aberkane A, Cuenca-Estrella M, Gomez-Lopez A et al (2002) Comparative evaluation of two different methods of inoculum preparation for antifungal susceptibility testing of filamentous fungi. J Antimicrob Chemother 50:719–722. https://doi.org/10.1093/JAC/DKF187
19. Abaya A, Serajazari M, Hsiang T (2021) Control of Fusarium head blight using the endophytic fungus, Simplicillium lamellicola, and its effect on the growth of Triticum aestivum. Biol Control 160:104684. https://doi.org/10.1016/J.BIOCONTROL.2021.104684
20. Møller CO d A, Freire L, Rosim RE et al (2021) Effect of lactic acid bacteria strains on the growth and aflatoxin production potential of aspergillus parasiticus, and their ability to bind aflatoxin B1, Ochratoxin A, and Zearalenone in vitro. Front Microbiol 12:655386. https://doi.org/10.3389/FMICB.2021.655386/BIBTEX
21. Wambacq E, Audenaert K, Höfte M et al (2018) Bacillus velezensis as antagonist towards Penicillium roqueforti s.l. in silage: in vitro and in vivo evaluation. J Appl Microbiol 125:986–996. https://doi.org/10.1111/JAM.13944
22. Alves SB, Moraes SA (1998) Quantificação de inóculo de patogênos de insetos. In: Controle microbiano de insetos
23. Racchi I, Scaramuzza N, Hidalgo A, Berni E (2020) Combined effect of water activity and pH on the growth of food-related ascospore-forming molds. Ann Microbiol 70:1–9. https://doi.org/10.1186/S13213-020-01612-6/TABLES/2
24. Voldřich M, Dobiáš J, Tichá L et al (2004) Resistance of vegetative cells and ascospores of heat resistant mould Talaromyces avellaneus to the high pressure treatment in apple juice. J Food Eng 61:541–543. https://doi.org/10.1016/S0260-8774(03)00224-3

25. Sant'Ana AS, Rosenthal A, Massaguer PR (2009) Heat resistance and the effects of continuous pasteurization on the inactivation of Byssochlamys fulva ascospores in clarified apple juice. J Appl Microbiol 107:197–209. https://doi.org/10.1111/j.1365-2672.2009.04195.x
26. Michener HD, King AD Jr (1974) Preparation of free heat-resistant Ascospores from Byssochlamys Asci. Appl Microbiol 27:671–673. https://doi.org/10.1128/AM.27.4.671-673.1974
27. Maneeboon T, Sangchote S, Hongprayoon R et al (2023) Modeling the thermal inactivation of Ascospores from heat-resistant molds in pineapple juice and evaluating disinfection efficiency of sodium hypochlorite and chlorine dioxide. Beverages 9:96. https://doi.org/10.3390/beverages9040096
28. Ausubel FM, Brent R, Kingston RE et al (1994) Current protocols in molecular biology. Wiley, Brooklyn
29. Sherman F (1991) [1] Getting started with yeast. Methods Enzymol 194:3–21. https://doi.org/10.1016/0076-6879(91)94004-V
30. Anapi GR, Aba RPM, Gabriel AA (2021) Screening for heat-resistant reference yeast isolate in orange juice. Food Microbiol 94:103639. https://doi.org/10.1016/J.FM.2020.103639
31. Treco DA, Winston F (2008) Growth and manipulation of yeast. Curr Protoc Mol Biol 82:13.2.1–13.2.12. https://doi.org/10.1002/0471142727.MB1302S82
32. Suiker IM, Arkesteijn GJA, Zeegers PJ, Wösten HAB (2021) Presence of Saccharomyces cerevisiae subsp. diastaticus in industry and nature and spoilage capacity of its vegetative cells and ascospores. Int J Food Microbiol 347:109173. https://doi.org/10.1016/J.IJFOODMICRO.2021.109173
33. Milani EA, Gardner RC, Silva FVM (2015) Thermal resistance of saccharomyces yeast ascospores in beers. Int J Food Microbiol 206:75–80. https://doi.org/10.1016/J.IJFOODMICRO.2015.04.002
34. Suiker IM, Kleijburg FEL, Wösten HAB (2023) Heat resistance acquirement of the spoilage yeast Saccharomyces diastaticus during heat exposure. J Food Prot 86:100020. https://doi.org/10.1016/J.JFP.2022.100020
35. Li L, Yang R, Zhao W (2021) The Effect of Pulsed Electric Fields (PEF) combined with temperature and natural preservatives on the quality and microbiological shelf-life of cantaloupe juice. Foods 10:2606. https://doi.org/10.3390/FOODS10112606
36. Kijpatanasilp I, Shiekh KA, Jafari S et al (2023) Microbial inhibition by UV radiation combined with Nisin and shelf-life extension of tangerine juice during refrigerated storage. Foods 12:2725. https://doi.org/10.3390/FOODS12142725/S1
37. Beuchat LR (1982) Thermal inactivation of yeasts in fruit juices supplemented with food preservatives and sucrose. J Food Sci 47:1679–1682. https://doi.org/10.1111/j.1365-2621.1982.tb05010.x
38. Rachon G, Raleigh CP, Pawlowsky K (2021) Heat resistance of yeast ascospores and their utilisation for the validation of pasteurisation processes for beers. J Inst Brew 127:149–159. https://doi.org/10.1002/JIB.646
39. Wickerham LJ (1951) Taxonomy of yeast. US Department of Agriculture, Washington DC. Technical Bulletin No. 1029., pp 1–59
40. García Carrillo M, Ferrario M, Guerrero S (2018) Effectiveness of UV-C light assisted by mild heat on Saccharomyces cerevisiae KE 162 inactivation in carrot-orange juice blend studied by flow cytometry and transmission electron microscopy. Food Microbiol 73:1–10. https://doi.org/10.1016/J.FM.2017.12.012
41. Salfinger Y, Tortorello ML (2015) Compendium of methods for the microbiological examination of foods, 15th edn. American Public Health Association, Washington, D.C.
42. Sabouraud R (1910) Les Teignes. Masson et Cie, Paris
43. Sabouraud R (1892) Contribution a l'etude de la trichophytie humaine. Etude clinique, microscopique et bacterioloqique sur la pluralité des trichophytons de l'homme. Ann Dermatol Syphilis 3:1061–1087
44. Put HMC, De Jong J, Sand FEMJ, Van Grinsven AM (1976) Heat resistance studies on yeast spp. causing spoilage in soft drinks. J Appl Bacteriol 40:135–152. https://doi.org/10.1111/J.1365-2672.1976.TB04162.X
45. Scott VN, Bernard DT (1985) Resistance of yeast to dry heat. J Food Sci 50:1754–1755. https://doi.org/10.1111/j.1365-2621.1985.tb10585.x
46. da Silva N, Taniwaki MH, Junqueira VCA et al (2019) Basic plate count techniques for enumeration of microorganisms. In: Microbiological examination methods of food and water, pp 25–46. https://doi.org/10.1201/9781315165011-3/BASIC-PLATE-COUNT-TECHNIQUES-

ENUMERATION-MICROORGANISMS-NEUSELY-DA-SILVA-MARTA-HIROMI-TANIWAKI-VAL

47. JOVE J of VE (2023) Basic methods in cellular and molecular biology: counting cells using Hemacytometer. https://app.jove.com/v/5048/using-a-hemacytometer-to-count-cells. Accessed 28 Oct 2024
48. Adams AM (2011) A comparative study of Ascospore formation by 43 yeast cultures. Can J Res 28f:413–416. https://doi.org/10.1139/CJR50F-037
49. Adams AM (2011) A convenient method of obtaining Ascospores from bakers' yeast. Can J Res 27c:179–189. https://doi.org/10.1139/cjr49c-015
50. McClary DO, Nulty WL, Miller GR (1959) Effect of potassium versus sodium in the sporulation of Saccharomyces. J Bacteriol 78:362–368. https://doi.org/10.1128/JB.78.3.362-368.1959
51. Put HMC, De Jong J (1982) Heat resistance studies of yeasts; vegetative cells versus ascospores: erythromycin inhibition of sporulation in Kluyveromyces and Saccharomyces species. J Appl Bacteriol 53:73–79. https://doi.org/10.1111/J.1365-2672.1982.TB04735.X
52. Put HMC, De Jong J (1982) The heat resistance of ascospores of four Saccharomyces spp. isolated from spoiled heat-processed soft drinks and fruit products. J Appl Bacteriol 52:235–243. https://doi.org/10.1111/J.1365-2672.1982.TB04845.X
53. Reddish A (1919) A substitute for beer wort as a bacteriological culture medium. J Bacteriol 3:6–7
54. Dutton S, Penn CW (1989) Biological attributes of colony-type variants of Candida albicans. J Gen Microbiol 135:3363–3372. https://doi.org/10.1099/00221287-135-12-3363/CITE/REFWORKS
55. Rapp M (1974) Indicator additions for differentiation of microorganisms on wort agar and malt-extract agar. Milchwissenschaft 29:341–344
56. Kleyn J (1954) A study of some environmental factors controlling sporulation of yeast using a new sporulation medium. Wallerstein Laboratories Communications 17:91
57. Beever RE, Bollard EG (1970) The nature of the stimulation of fungal growth by potato extract. Microbiology (N Y) 60:273–279. https://doi.org/10.1099/00221287-60-2-273
58. Pomper S, Daniels KM, McKee DW (1954) Genetic analysis of polyploid yeast. Genetics 39:343–355. https://doi.org/10.1093/genetics/39.3.343
59. Splittstoesser DF, Leasor SB, Swanson KMJ (1986) Effect of food composition on the heat resistance of yeast Ascospores. J Food Sci 51:1265–1267. https://doi.org/10.1111/j.1365-2621.1986.tb13101.x

# Chapter 8

# Determination of Heat Resistance of Microorganisms

**Roberto Pessanha da Silva Pires and Matheus S. Mourão**

**Abstract**

The text emphasizes the importance of microbiological control in the food industry, particularly regarding the thermal resistance of microorganisms during food processing. Food contamination poses a significant public health issue, with foodborne diseases affecting 1 in 10 people globally each year and resulting in the loss of millions of healthy years. The resistance of microorganisms, particularly spores, to thermal and nonthermal treatments is a significant concern. The text describes methods for determining the isothermal and nonisothermal heat resistance of microorganisms, including defining thermal death points and using D- and z-values to evaluate the effectiveness of thermal treatments.

**Key words** Isothermal, Nonisothermal treatments, TDT

## 1 Introduction

Food contamination represents a significant public health issue. Various forms of contamination during food processing have driven the development of new technologies for microbial control in industrial processes. The World Health Organization [1] estimates that 1 in 10 people worldwide falls ill each year due to foodborne diseases, resulting in approximately 33 million healthy life years lost. To ensure the safe development of food, it is crucial to control factors such as process conditions, product properties, hygiene, and storage conditions. Methods for inactivating undesirable microorganisms depend on factors such as water activity (aw), pH, initial microorganism levels, growth characteristics, and inactivation processes, as well as storage conditions of time and temperature [2–4].

Among the microorganisms with the potential to contaminate food during processing, *Listeria monocytogenes*, *Salmonella*, *Escherichia coli*, *Campylobacter*, and *Staphylococcus aureus* are particularly notable, along with the species of *Lactobacillus* and *Pseudomonas*. Spore-forming microorganisms such as *Bacillus cereus*, *Clostridium perfringens*, and *Clostridium botulinum* are frequently associated with contamination events [2, 5]. The resistance of spores,

Anderson S. Sant'Ana (ed.), *Heat Resistant Microorganisms in Food Safety and Spoilage*, Methods and Protocols in Food Science, https://doi.org/10.1007/978-1-0716-5280-0_8,

vegetative cells, enteric strains of various species, and microorganisms from different environments varies with each analysis, leading to different results depending on the processing method [2].

The primary concern regarding spores in food processing is their high resistance to thermal and nonthermal treatments [6]. Several factors can facilitate germination, including the presence of nucleosides, sugars, and amino acids [7]. In this sense, den Besten et al. [2] often link these contaminants to suboptimal processing conditions and to the need for precise knowledge of microorganism inactivation kinetics. Scientists have consequently developed various techniques to improve the inactivation of microorganisms, particularly by enhancing thermal processing controls in industrial applications.

The process of heat microbial control resistance requires a specific thermal death point, which describes the sensitivity of bacteria to the treatment application. This concept relates to the time needed to inactivate the suspension at a constant temperature [8]. Researchers use D-values and z-values to evaluate these characteristics and apply the analyses. The D-values are used to estimate the time required for a 1-log colony-forming unit reduction or a 90% reduction in the microbial population. For bacteria, this factor is influenced by the processing method used, the type and strain of bacterium tested, the heating matrix, and the temperature applied [9]. The z-values are the temperatures required to reduce the D-value by a factor of 10. They are more critical for evaluating bacterial resistance to thermal treatment within a specific temperature range [10].

Determining heat resistance in pathogenic or spoilage microorganisms is essential for maintaining food safety because it describes the time–temperature binomial for microbial inactivation. *Salmonella* and *Clostridium botulinum* are target microorganisms for several food matrices, and their D- and z-values help to determine standard inactivation methods for these matrices. While isothermal assays provide baseline parameters, non-isothermal models simulate industrial histories by integrating momentary inactivation rates over time [11–13].

## 2 Isothermal Methods for Determining Heat Resistance

### 2.1 Objective

To establish *D-values* and *z-values*, which are *thermal resistance parameters* of foodborne microorganisms under *isothermal heating conditions*. These parameters are determined in liquid or semisolid food matrices (e.g., milk, juice, fruit puree) or in model systems (e.g., 0.85% (w/v) NaCl solution, 0.1 M phosphate buffer) using TDT tubes, TDT cans, capillary tubes, and microtubes.

## 2.2 Principle

Food matrices (or model systems) are treated with microbial suspensions at a *constant temperature*. At the initial point and at each predetermined time interval, aliquots are taken and counted to determine the microorganism concentration in the matrix *(log CFU/mL) over time*, and a graph is plotted. *D-value* is the time the treatment takes for the microorganism concentration to decay by 1 log (90%) at this constant temperature. The *z-value* is the temperature increase required for the D-value to decrease by 10 times (90%) [14].

## 2.3 Safety Considerations

- All treatments and analyses must be performed under biosafety conditions, depending on the microorganisms used.
- When using heat, wear heat-resistant gloves and safety glasses.
- After use, autoclave (121 °C for 15 min) all used materials before disposal or storage.

## 2.4 Microorganisms and Culture Preparation

### 2.4.1 Microorganisms

- Use pathogenic or spoilage microorganisms (e.g., *Listeria monocytogenes*, *Salmonella*, and *Bacillus* spores) that are well characterized and obtained from standardized collections (e.g., ATCC, NCTC, Embrapa, Fiocruz).
- When using spores, verify that at least 90% have sporulated before incubation.

### 2.4.2 Culture Preparation

| | |
|---|---|
| *STEP 1* | Inoculate the microorganism into the broth most suitable for its growth (e.g., BHI, TSB). |
| *STEP 2* | Incubate the cultures under optimal conditions until they reach the late exponential or early stationary growth phase. |
| *STEP 3* | Collect the cells by centrifugation (optional) and then resuspend them in a food matrix (e.g., milk, juice) or in a model system (e.g., NaCl solution, phosphate buffer). |
| *STEP 4* | Concentrate the inoculum until $10^7$ to $10^9$ CFU/mL. |

## 2.5 Heating Media and Test Systems

### 2.5.1 Heating Media

Experiments can be carried out using standard models as heating media, such as phosphate buffer [15] or saline solution [16], or food matrices, such as emulsion-type buffalo meat [15], juices [17], milk [18], and infant formula [19]. Using standard models, pH can be controlled; in food matrices, many parameters can be studied, such as pH, fat content, and water activity.

### 2.5.2 Isothermal Systems

Depending on the required aliquot volume or the type of heating media used, different containers can be used for the study.

- Capillary tubes (e.g., 50–100 μL), as used by [20, 21].
- TDT glass tubes (sealed, thin-walled, 1–2 mL), as used by [22].

- TDT metal cans (used for larger volumes or viscous foods), as used by [23].
- Microcentrifuge tubes (less ideal; only if validated), as used by [24].

### 2.6 Equipment

- Water or oil bath (±0.1 °C accuracy) for calibrated heating.
- Thermometer or thermocouple, calibrated and preferably the same for every test.
- Ice–water bath, for rapid quenching.
- Vortex mixer, for homogenizing the samples during serial dilutions.
- Pipettes and sterile tips, for serial dilutions.
- Incubator, for microbial enumeration.

### 2.7 Experimental Procedure

#### 2.7.1 Filling and Sealing

| | |
|---|---|
| *STEP 1* | Fill the containers with the required volume of inoculum: Capillary tubes (50–100 μL), TDT glass tubes (1–2 mL), TDT metal cans (as needed). |
| *STEP 2* | Eliminate the air bubbles. |
| *STEP 3* | Close the tubes or capillaries by flame-sealing or using a suitable sealing material. |

#### 2.7.2 Heat Treatment (Isothermal Exposure)

| | |
|---|---|
| *STEP 1* | Preheat the water or oil bath to the selected temperature (e.g., 50, 55, or 60 °C). |
| *STEP 2* | Introduce the test units into the bath only after confirming that the temperature is stable. |
| *STEP 3* | Begin recording exposure time when the coldest location in the sample achieves the set temperature (the come-up time should be established in prior trials). |
| *STEP 4* | Collect the samples at scheduled time points (for instance, 0, 1, 2, 4, 6, and 8 min). |
| *STEP 5* | Transfer the samples immediately to an ice–water bath to arrest thermal inactivation rapidly. |

### 2.8 Microbial Enumeration

| | |
|---|---|
| *STEP 1* | Dilute 10 mL of the liquid food matrix (e.g., infant formula) with 90 mL of 0.85% (w/v) NaCl solution and then homogenize for 5 min. After homogenization, this solution will be considered as sample $10^{-1}$. |
| *STEP 2* | Dilute 1 mL of *sample* $10^{-1}$ in 9 mL of 0.85% (w/v) NaCl solution and homogenize the solution to constitute *sample* $10^{-2}$. |
| *STEP 3* | Thus, successively make serial dilutions by diluting 1 mL of the previous sample in 9 mL of 0.85% (w/v) NaCl solution, continuing until *sample* $10^{-7}$ is generated. |
| *STEP 4* | Spread plate 0.1 mL of each sample ($10^{-1}$ to $10^{-7}$) on three petri plates with appropriate selective or nonselective agar (e.g., bismuth sulfite agar) and incubate for 24 h at 37 °C to verify the growth of microorganisms (e.g., *Salmonella*, one of the primary microorganisms causing foodborne illnesses). |
| *STEP 5* | For plate counting, select a sample ($10^{-1}$ to $10^{-7}$) in which the petri plates contain 10 to 100 black metallic colonies each. |
| *STEP 6* | To determine the concentration of the microorganism, it must multiply the average of colonies (in the three plates of the same dilution) by $10^2$ (because of the planned volume) and the opposite of dilution (if the chosen dilution was $10^{-3}$, it must be multiplied by $10^3$). This calculation is crucial for accurately determining the concentration of the microorganism in the original sample. In an example where the average bacterial count was 53 in *sample* $10^{-3}$ plates, the original sample concentration is $53 \times 10^2 \times 10^3 = 5.3 \times 10^6$ CFU/mL. |

### 2.9 Data Analysis

#### 2.9.1 Survival Curves

| | |
|---|---|
| *STEP 1* | Plot the logarithm (base 10) of the surviving cells concentration as a function of heating time. |
| *STEP 2* | Check whether the inactivation curve is linear, confirming first-order kinetic behavior. |

#### 2.9.2 D-Value Calculation

- $D_T$ is calculated as −1 divided by the slope, which is determined from the linear regression of the survival curve.

#### 2.9.3 Z-Value (Optional)

| | |
|---|---|
| *STEP 1* | Obtain D-values at several different treatment temperatures for the same microorganism and food matrix. |
| *STEP 2* | Plot the logarithm (base 10) of the D-values as a function of temperature. |
| *STEP 3* | The z-value is the temperature increase required for the D-value to decay by 1 log (90%). |

### 2.10 Quality Control and Validation

| | |
|---|---|
| *STEP 1* | Include non-heated samples as controls. |
| *STEP 2* | Carry out the experiments with both biological and technical replicates. |
| *STEP 3* | Verify that the bath temperature remains stable throughout all runs. |
| *STEP 4* | Ensure that no post-treatment recovery of microorganisms occurs unless such recovery is an explicit objective of the study. |

### 2.11 Reporting Results

**Results Should Include**

- The pathogenic or spoilage microorganism used and its strain
- Culture conditions and sample preparation procedures
- Type of heating system employed (e.g., TDT tube, capillary system, and related configurations)
- Composition of the food or model matrix and its pH value
- Processing temperature(s) evaluated
- D-values reported as mean ± standard deviation
- Any deviations from linear behavior observed in the inactivation curves

### 2.12 Notes and Limitations

- Capillary tubes allow high-speed heat transfer, but their use is essentially restricted to liquid samples.
- TDT cans are better suited to viscous or particulate food products, though they require larger corrections due to longer come-up times.
- When survival curves deviate from the log-linear behavior, alternative kinetic models may be needed to describe inactivation.

## 3 Non-isothermal Methods for Determining Heat Resistance

### 3.1 Objective

To establish equivalent lethality, kinetic parameters, or *D-values* and *z-values*, which are *thermal resistance parameters* of foodborne microorganisms under *non-isothermal (dynamic) heating conditions.* These parameters are determined in food matrices (e.g., milk, juice, fruit puree) or in model systems (e.g., 0.85% (w/v) NaCl solution, 0.1 M phosphate buffer) using TDT tubes, TDT cans, capillary tubes, and microtubes, which represent real industrial treatments.

### 3.2 Principle

Food matrices (or model systems) are treated with microbial suspensions at *continuously changing temperatures.* At the initial point and at each predetermined time interval, aliquots are taken and counted to determine the microorganism concentration in the matrix (log CFU/mL) over time, and a graph is plotted. The sample temperature must be recorded at each time. GinaFit (Excel® add-in) is used to generate thermal inactivation kinetics (e.g., Bigelow or alternative models) [25]. It is necessary to compare the measured log reduction with the predicted lethality in order to estimate or confirm the thermal resistance parameters.

### 3.3 Safety Considerations

- All treatments and analyses must be performed under biosafety conditions, depending on the microorganisms used.
- When using heat, wear heat-resistant gloves and safety glasses.
- After use, autoclave (121 °C for 15 min) all used materials before disposal or storage.

### 3.4 Microorganisms and Culture Preparation

#### 3.4.1 Microorganisms

- Use pathogenic or spoilage microorganisms (e.g., *Listeria monocytogenes*, *Salmonella*, and *Bacillus* spores) that are well characterized and obtained from standardized collections (e.g., ATCC, NCTC, Embrapa, Fiocruz).
- When using spores, verify that at least 90% have sporulated before incubation.

#### 3.4.2 Culture Preparation

| | |
|---|---|
| *STEP 1* | Inoculate the microorganism into the broth most suitable for its growth (e.g., BHI, TSB). |
| *STEP 2* | Incubate the cultures under optimal conditions until they reach the late exponential or early stationary growth phase. |
| *STEP 3* | Collect the cells by centrifugation (optional) and then resuspend them in a food matrix (e.g., milk, juice) or in a model system (e.g., NaCl solution, phosphate buffer). |
| *STEP 4* | Concentrate the inoculum until $10^7$ to $10^9$ CFU/mL or g. |

### 3.5 Heating Media and Test Systems

#### 3.5.1 Heating Media

Experiments can be carried out using standard models as heating media, such as phosphate buffer [15] or saline solution [16], or food matrices, such as emulsion-type buffalo meat [15], juices [17], milk [18], and infant formula [19]. Using standard models, pH can be controlled; in food matrices, many parameters can be studied, such as pH, fat content, solids, and water activity.

#### 3.5.2 Containers

Depending on the required aliquot volume or the type of heating media used, different containers can be used for the study:

- Capillary tubes (e.g., 50–100 μL), as used by [20, 21].
- TDT glass tubes (sealed, thin-walled, 1–2 mL), as used by [22].
- TDT cans or retort pouches (used for larger volumes or particulate/viscous foods), as used by [23].
- Pilot-scale vessels (if applicable).

### 3.6 Equipment

- Water bath, oil bath, or retort (±0.1 °C accuracy) that can be programmable for controlling temperature increases and decreases.
- Thermometer or thermocouple loggers that provide a quick response.
- Ice-water bath, for rapid quenching.
- System for recording and storing time–temperature data.
- Vortex mixer, for homogenizing the samples during serial dilutions.
- Pipettes and sterile tips, for serial dilutions.
- Incubator, for microbial enumeration.

### 3.7 Experimental Procedure

#### 3.7.1 Filling and Sealing

| | |
|---|---|
| *STEP 1* | Fill the containers with the required volume of inoculum: Capillary tubes (50–100 μL), TDT glass tubes (1–2 mL), TDT metal cans (as needed). |
| *STEP 2* | Eliminate the air bubbles. |
| *STEP 3* | Close the tubes or capillaries by flame-sealing or using a suitable sealing material. |

#### 3.7.2 Establishment of Nonisothermal Profile

| | |
|---|---|
| *STEP 1* | Choose the heating profile to determine when the sample temperature increases and decreases (e.g., 25 → 65 °C at 5 °C/min, hold 2 min, cool). |
| *STEP 2* | Set the heating equipment to follow the desired temperature profile. |
| *STEP 3* | Validate the temperature profile by placing thermocouples at the coldest location in the product and confirming that it follows the intended pattern. |

#### 3.7.3 Heat Treatment (Nonisothermal Exposure)

| | |
|---|---|
| *STEP 1* | Load the samples into the heating apparatus. |
| *STEP 2* | Continuously record the temperature during both the heating and cooling stages. |
| *STEP 3* | Collect the samples at scheduled time points (for instance, 0, 1, 2, 4, 6, and 8 min). |
| *STEP 4* | Transfer the samples immediately to an ice–water bath to arrest thermal inactivation rapidly. |

### 3.8 Enumeration

| | |
|---|---|
| *STEP 1* | Dilute 10 mL of the liquid food matrix (e.g., infant formula) with 90 mL of 0.85% (w/v) NaCl solution and then homogenize for 5 min. After homogenization, this solution will be considered as *sample* $10^{-1}$. |
| *STEP 2* | Dilute 1 mL of *sample* $10^{-1}$ in 9 mL of 0.85% (w/v) NaCl solution and homogenize the solution to constitute *sample* $10^{-2}$. |
| *STEP 3* | Thus, successively make serial dilutions by diluting 1 mL of the previous sample in 9 mL of 0.85% (w/v) NaCl solution, continuing until *sample* $10^{-7}$ is generated. |
| *STEP 4* | Spread plate 0.1 mL of each sample ($10^{-1}$ to $10^{-7}$) on three petri plates with appropriate selective or nonselective agar (e.g., bismuth sulfite agar) and incubate for 24 h at 37 °C to verify the growth of microorganisms (e.g., *Salmonella*, one of the primary microorganisms causing foodborne illnesses). |
| *STEP 5* | For plate counting, select a sample ($10^{-1}$ to $10^{-7}$) in which the Petri plates contain 10 to 100 black metallic colonies each. |
| *STEP 6* | To determine the concentration of the microorganism, multiply the average of colonies (in the three plates of the same dilution) by $10^2$ (because of the planned volume) and the opposite of dilution (if the chosen dilution was $10^{-3}$, it must be multiplied by $10^3$). This calculation is crucial for accurately determining the concentration of the microorganism in the original sample. In an example where the average bacterial count was 53 in *sample* $10^{-3}$ plates, the original sample concentration is $53 \times 10^2 \times 10^3 = 5.3 \times 10^6$ CFU/mL. |

### 3.9 Data Treatment and Kinetic Analysis

#### 3.9.1 Time–Temperature History

| | |
|---|---|
| *STEP 1* | For each sample, record the full temperature-time profile. |
| *STEP 2* | Adjust it to account for the come-up and cooling periods. |

*3.9.2 Lethality Calculation (F) (Bigelow Approach)*

Calculate the equivalent time at a reference temperature ($T_{ref}$):

$$F_{\{T_{\{ref\}}\}} = \int_0^t 10\frac{T_{(t)} - T_{\{ref\}}}{z}, dt$$

where $F_{\{T_{\{ref\}}\}}$ is the equivalent processing time at the reference temperature $T_{\{ref\}}$; $T_{(t)}$ is the product temperature as a function of time; $T_{\{ref\}}$ is the reference temperature; z is the temperature change (°C) required to achieve a tenfold change in the decimal reduction time (D-value); and t is the processing time.

*3.9.3 Estimation of Thermal Resistance*

*STEP 1* Compare observed log reductions with predicted lethality to verify that measured microbial inactivation matches or exceeds the inactivation expected from the calculated F-value for the process.

*STEP 2* Estimate Dref and z-values by fitting thermal inactivation data with regression or optimization procedures applied to the Bigelow-type relationship between log D and temperature.

*STEP 3* Alternatively, fit non-linear or Weibull models when survival curves depart from log-linearity.

### 3.10 Model Validation

*STEP 1* Use independently determined isothermal D- and z-values to predict non-isothermal inactivation from the recorded T (t) profiles, by integrating the time–temperature history into the chosen kinetic model.

*STEP 2* Compare predicted vs. observed log reductions to assess how well the lethality model reproduces experimental inactivation, and to identify any systematic under- or overestimation of microbial kill.

*STEP 3* Assess goodness-of-fit using $R^2$ to quantify the proportion of variance explained and RMSE to measure the average deviation between observed and predicted log reductions, where higher $R^2$ and lower RMSE indicate better agreement.

### 3.11 Quality Control

*STEP 1* Include non-heated samples as controls.

*STEP 2* Carry out the experiments with both biological and technical replicates.

*STEP 3* Verify that the bath temperature remains stable (accuracy and repeatability) throughout all runs (triplicate).

*STEP 4* Evaluate biological variability by quantifying strain-to-strain and population-to-population differences in thermal resistance (e.g., distributions of D- and z-values) and considering how prior stress and adaptation modify these responses.

### 3.12 Reporting Results

**Report**

- The pathogenic or spoilage microorganism used and its strain.
- Culture conditions and sample preparation procedures.
- Type of heating system employed and its volume (e.g., TDT tube, capillary system, and related configurations).
- Composition of the food or model matrix and its pH value.
- Heating profile (Temperature x time graphic).
- Observed log reductions (log CFU/mL or g).
- Model assumptions and limitations.

### 3.13 Notes and Limitations

- Nonisothermal methods better mimic industrial heating profiles and process dynamics, but their results depend more on the chosen mathematical model and its underlying assumptions.
- Accurate temperature measurement is critical to ensure that thermal processes achieve the intended microbial inactivation, product safety, and regulatory compliance.
- Model choice strongly influences parameter estimation, since different kinetic structures can fit the same data yet yield markedly different parameter estimates and uncertainties.

### 3.14 Typical Applications

- Validation of pasteurization or sterilization processes, including confirmation that the applied thermal treatment achieves the target microbial log reduction under defined time–temperature conditions.
- Comparison of thermal resistance across food matrices involves evaluating how the D- and z-values of target microorganisms change when heated in different food systems (e.g., high-fat, low-moisture, liquid, or structured products).
- Evaluation of dynamic heating in novel technologies focuses on how time–temperature profiles develop and change within foods when using advanced methods such as ohmic, microwave, and radio-frequency heating.

## 4 Notes

The control of microorganisms' thermal resistance is crucial to ensuring food safety, especially in industrial processes designed to inactivate pathogens. The described techniques are essential for evaluating the effectiveness of applied treatments and monitoring microbial resistance during food processing. Understanding the D- and z-values, which help determine the effectiveness of thermal processes, is essential to optimizing processing parameters and minimizing contamination risks. Advancements in microbial inactivation techniques, particularly controlled thermal treatments, can enhance food safety and reduce the incidence of foodborne diseases. In addition to isothermal tests, nonisothermal tests are essential because they more closely resemble industrial conditions, enabling the determination of more realistic microbial inactivation parameters.

## References

1. WHO. World Health Organization (2022) Foodborne disease estimates. https://www.who.int/data/gho/data/themes/food-safety
2. den Besten HMW, Wells-Bennik MHJ, Zwietering MH (2018) Natural diversity in heat resistance of bacteria and bacterial spores: impact on food safety and quality. Annu Rev Food Sci Technol. 9:383–410. https://doi.org/10.1146/annurev-food-030117-012808
3. Membré J-M (2016) Microbiological risk assessments in the food industry. Food hygiene and toxicology in ready-to-eat foods. Elsevier:337–350. https://doi.org/10.1016/B978-0-12-801916-0.00019-4
4. Zwietering MH (2015) Risk assessment and risk management for safe foods: assessment needs inclusion of variability and uncertainty, management needs discrete decisions. Int J Food Microbiol. 213:118–123. https://doi.org/10.1016/j.ijfoodmicro.2015.03.032
5. Peck MW, van Vliet AH (2016) Impact of *Clostridium botulinum* genomic diversity on food safety. Curr Opin Food Sci. 10:52–59. https://doi.org/10.1016/j.cofs.2016.09.006
6. Black EP, Setlow P, Hocking AD, Stewart CM, Kelly AL, Hoover DG (2007) Response of spores to high-pressure processing. Compr Rev Food Sci Food Saf. 6:103–119. https://doi.org/10.1111/j.1541-4337.2007.00021.x
7. Krawczyk AO, de Jong A, Omony J, Holsappel S, Wells-Bennik MHJ, Kuipers OP et al (2017) Spore heat activation requirements and germination responses correlate with sequences of germinant receptors and with the presence of a specific *spoVA* 2mob operon in foodborne strains of *Bacillus subtilis*. Appl Environ Microbiol. 83(7):e03122-16. https://doi.org/10.1128/AEM.03122-16
8. Tischer RG, Hurwicz H (1954) Thermal characteristics of bacterial populations a. J Food Sci. 19:80–91. https://doi.org/10.1111/j.1365-2621.1954.tb17425.x
9. Zeki B (2009) Food process engineering and technology. Elsevier/Academic
10. Casadei MA, de Matos RE, Harrison G (1998) Heat resistance of *Listeria monocytogenes* in dairy products as affected by the growth medium. J Appl Microbiol. 84:234–239. https://doi.org/10.1046/j.1365-2672.1998.00334.x
11. Garre A, González-Tejedor G, Peñalver-Soto JL, Fernández PS, Egea JA (2018) Optimal characterization of thermal microbial inactivation simulating non-isothermal processes. Food Res Int 107:267–274
12. Soni A, Bremer P, Brightwell G (2022) A comprehensive review of variability in the thermal resistance (D-values) of food-borne pathogens—A challenge for thermal validation trials. Foods 11(24):4117
13. Dora-Liyana AL, Mahyudin NA, Ismail-Fitry MR, Ahmad Zainuri MD (2023) Thermal inactivation D- and z-values of *Salmonella* Enteritidis and *Salmonella* Typhimurium in whole muscle beef. Malay Appl Biol 52(6): 119–126. https://doi.org/10.55230/mabjournal.v52i6.2776
14. Peñalver-Soto JL, Garre A, Esnoz A, Fernández PS, Egea JA (2019) Guidelines for the design of (optimal) isothermal inactivation experiments. Food Res Int 126:108714
15. Devadason IP, Anjaneyulu ASR, Mendirtta SK, Murthy TRK (2014) Determination of thermal process schedule for emulsion type buffalo meat block in retort pouch. J Food Sci Technol 51(11):3482–3487
16. Kim C, Alrefaei R, Bushlaibi M, Ndegwa E, Kaseloo P, Wynn C (2019) Influence of growth temperature on thermal tolerance of leading foodborne pathogens. Food Sci Nutr 7(12): 4027–4036
17. Pihen C, Mani-López E, López-Malo A, Ramírez-Corona N (2024) Modeling non-isothermal pathogen inactivation during liquid foods processing by pulsed light technology. J Food Process Eng 47(1):e14521
18. Lau SK, Subbiah J (2020) TDT Sandwich: an open source dry heat system for characterizing

the thermal resistance of microorganisms. HardwareX 8:e00114

19. Pires RPS, Guimarães JT, Barros CP, Balthazar CF, Chincha AIA, Freitas MQ et al (2021) Ohmic heating increases inactivation and morphological changes of *Salmonella* sp. and the formation of bioactive compounds in infant formula. Food Microbiol. 97:103737. https://doi.org/10.1016/j.fm.2021.103737
20. Alvarenga VO, Campagnollo FB, Pia AK, Conceição DA, Abud Y, Sant'Anna C et al (2018) Quantifying the responses of three *Bacillus cereus* strains in isothermal conditions and during spray drying of different carrier agents. Front Microbiol *9*:1113
21. Rana YS, Chen L, Balasubramaniam VM, Snyder AB (2022) Superheated steam effectively inactivates diverse microbial targets despite mediating effects from food matrices in bench-scale assessments. Int J Food Microbiol *378*:109838
22. Büchner C, Thomas S, Jaros D, Rohm H (2012) Fast-responding thermal-death-time tubes for the determination of thermal bacteria inactivation. Eng Life Sci *12*(1):109–112
23. Chung HJ, Birla SL, Tang J (2008) Performance evaluation of aluminum test cell designed for determining the heat resistance of bacterial spores in foods. LWT-Food Sci Technol *41*(8):1351–1359
24. Oscar TP (2017) Modeling the effect of inoculum size on the thermal inactivation of *Salmonella* Typhimurium to elimination in ground chicken thigh meat. Am J Food Sci Technol *5*(4):135–142
25. Geeraerd AH, Valdramidis VP, Van Impe JF (2005) GInaFiT, a freeware tool to assess non-log-linear microbial survivor curves. Int J Food Microbiol 102(1):95–105

# Chapter 9

# Estimating Inactivation Kinetic Parameters of Microorganisms Log-Linear Kinetics

**Mahmoud Yolmeh and Matheus Péricles Silva Láscaris**

**Abstract**

Microbial inactivation kinetic models play a crucial role in food safety, public health, and environmental microbiology by providing predictive information for the design of thermal processes. In these models, the focus generally lies on the rate of microbial inactivation. The fundamental equations and key concepts of log-linear models are discussed, highlighting their applications and limitations, and statistical tools for evaluating model accuracy, such as $R^2$, RMSE, $A_f$, and $B_f$, are presented. In addition, the calculation methods for D- and Z-values are detailed.

**Key words** Microbial inactivation, Log-linear model, Isothermal condition, Model accuracy

## 1 Introduction

Microorganism inactivation is critical in food safety, healthcare, and environmental microbiology. Microbial inactivation reduces or eliminates viable microbial populations through physical, chemical, or biological treatments. This process is fundamental to food preservation techniques such as pasteurization, sterilization, and disinfection [1].

Microbial inactivation kinetic models provide a mathematical framework to understand and predict how microbial populations decline over time. These models are indispensable for designing and optimizing thermal processes, validating sterilization procedures, and ensuring regulatory compliance (*see* **Note 1**). Among the various kinetic models, log-linear models are widely used because they offer simplicity, accuracy, and adaptability to many microbial inactivation scenarios [2].

This chapter reviews the basics of log-linear models, their types, and their applications. Special attention is given to preparing samples, experimental methodologies, and parameter calculation (*see* **Note 5**).

Anderson S. Sant'Ana (ed.), *Heat Resistant Microorganisms in Food Safety and Spoilage*, Methods and Protocols in Food Science, https://doi.org/10.1007/978-1-0716-5280-0_9,

**Definition of Kinetic Models and Their Importance in Microbiology**

Microbial inactivation kinetic models describe the rate at which microbial populations change over time when exposed to inactivating agents, such as heat. They provide a quantitative understanding of microbial death, enabling predictions about how long it takes to achieve desired safety or spoilage control under specific conditions by making a relationship between treatment conditions (temperature and time) and microbial count reduction [2].

- *Predictive capability*: To estimate microbial survival or inactivation during processes like pasteurization, sterilization, and cooking
- *Process design*: To guide the selection of time–temperature combinations that maximize microbial inactivation while preserving food quality (*see* **Note 3**)
- *Regulatory compliance*: To provide evidence to meet safety standards set by organizations such as the FDA, EFSA, and WHO

**Log-Linear Models**

The log-linear model is the simplest and most widely used inactivation model. The first and most important feature required to use this model is to assume that microbial death occurs at a constant proportion over time (*see* **Note 4**). The log-linear model is derived from the principles of the first-order model (Eq. 1) as follows:

$$\frac{dN}{dt} = -kN \tag{1}$$

where $c$, $t$, and $k$ represent the microbial population, heating time in isothermal conditions (min), and rate constant ($\text{min}^{-1}$), respectively. By solving this differential equation, the first-order model predicts the number of microorganisms at any time ($t$):

$$N = N_0\, e^{-kt} \tag{2}$$

where $N_0$ is the initial number of microorganisms. This exponential model is the basis for microbial inactivation studies. However, for easy analysis of data, particularly in microbial inactivation kinetics, the first-order model is often transformed into a log-linear form by taking the logarithm of both sides of the primary exponential model:

$$\log_{10}(N) = \log_{10}(N_0) - \frac{k}{\ln(10)}\, t \tag{3}$$

And in a simpler form:

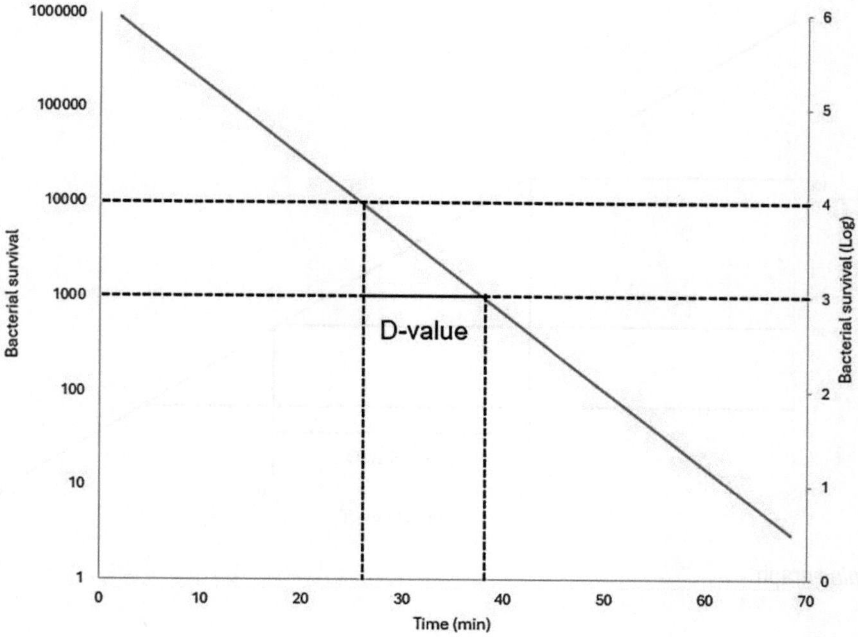


**Fig. 1** *D-value* graph

$$\log_{10}(N) = \log_{10}(N_0) - D\,t \tag{4}$$

where $D$ is the decimal reduction time (in min) or *D-value* ($\frac{1}{k\ \ln(10)}$), which is defined in simpler terms as "the length of time it takes for a population to decrease by 90 percent (one-log)." This parameter indicates the heat sensitivity of microorganisms at a given temperature so that lower *D-values* indicate the greater sensitivity of the microorganism to the applied thermal conditions (*see* **Note 2**). This equation presents a straight-line relationship when $\log_{10}(N)$ is plotted over time ($t$), and for this reason, the term is named *the log-linear model.* If it is necessary to determine the *D-value* from a graph, the slope of the line indicates it (Fig. 1) [3].

When log *D-value*s are plotted against temperature, they typically form a straight line, known as the thermal death time curve. The *z-value* is derived from this curve and indicates the temperature increase needed to reduce the *D-value* by 90 percent (one-log) (Fig. 2):

$$z = \frac{T_2 - T_1}{\log D_{T1} - \log D_{T2}}$$

where $T_1$ and $T_2$ are two different temperatures applied, and $D_T$ is the *D-value* measured at temperature T [4].

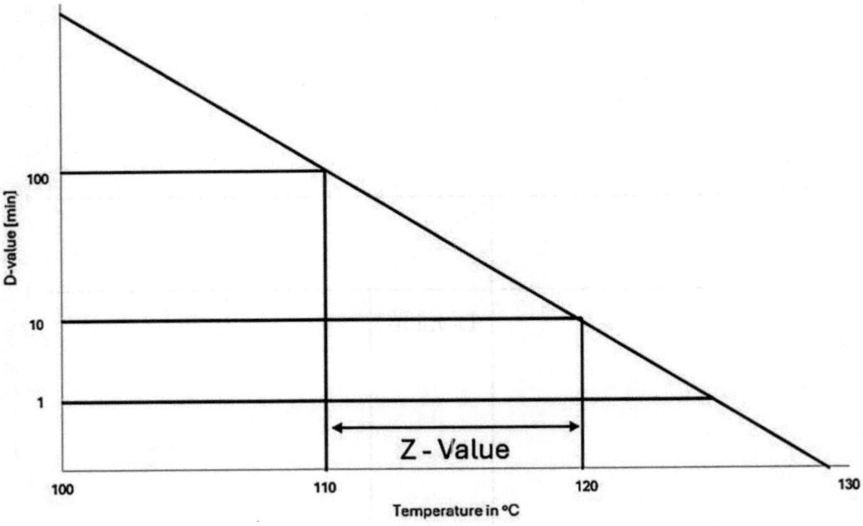


**Fig. 2** *z-value* graph

## Different Types of Log-Linear Models

Log-Linear Model

- *Concept*: Assumes an exponential decline in microbial populations due to constant inactivation rates
- *Equation*:

$$\log_{10}(N) = \log_{10}(N_0) - D\,t \tag{5}$$

where $N_0$ and $N$ are the initial microbial population and microbial population at time t, respectively.

Applications

- Widely used for describing microbial death during pasteurization and sterilization
- Suitable for processes where all microorganisms exhibit similar sensitivity

The log-linear kinetic model has been used in several studies to investigate the effect of various factors on the survival of bacteria in food microbiology and sterilization processes. The high use of this model is due to its ease of use because of the uniform consideration of temperature during the thermal process.

**Log-Linear Shoulder Model**

- *Concept*: The log-linear shoulder model accounts for a lag phase, known as the "shoulder," at the beginning of inactivation. This initial delay represents a period where the microbial population is resistant to the stressor before a log-linear decline begins.
- *Equation*:

$$\log N_t = \log N_0 + k(S - t).\log(e) - \log\left[1 + \left(e^{k.s} - 1\right).e^{-k.t}\right] \quad (6)$$

where $N_0$, $N_t$, $t$, $k$, and $S$ represent the initial microbial population, microbial population at time t, treatment time, microbial inactivation rate, and shoulder length, respectively.

- *Characteristics*: In this model, the inactivation curve shows a "shoulder" at the beginning, where there is little to no reduction in microbial population initially, followed by a log-linear decline. This fact makes this model more accurate for certain microbial populations that adapt or require time to begin inactivation.
- *Applications*: Processes involving sublethal stress, where microorganisms require time to adapt before inactivation [5].

**Log-Linear Shoulder and Tail Model**

- *Concept*: This model incorporates both an initial shoulder (delay) and a tailing effect at the end of the inactivation process. The "tail" represents a phase where a portion of the microbial population becomes highly resistant, leading to a plateau where few or no organisms are inactivated, even with prolonged exposure.
- *Equation*:

$$\log N_t = \log(N_0 - N_r) + k(S - t).\log(e) - \log\left[1 + \left(e^{k.s} - 1\right).e^{-k.t}\right] + N_r \quad (7)$$

where $N_0$, $N_r$, $N_t$, $t$, $k$, and $S$ represent the initial microbial population, residual microbial population, microbial population at time t, treatment time, microbial inactivation rate, and shoulder length, respectively.

- *Characteristics*: The curve starts with a shoulder, followed by a log-linear decline, and finally exhibits a tail at the end, where the inactivation rate decreases, resulting in a more asymptotic approach. This model is the most comprehensive of the three, addressing both the initial delay and the tailing effect, making it more suitable for cases where microbial populations show varied responses, including high resistance in some subpopulations.
- *Applications*: Used when experimental data indicate both an initial lag and a tailing effect, such as in heat-resistant spore inactivation [5, 6].

**Comparative Summary**

- *Log-Linear Model*: Assumes a constant, exponential inactivation rate with no initial shoulder or tail
- *Log-Linear Shoulder Model*: Includes an initial shoulder period where inactivation starts slowly and then follows a log-linear decline
- *Log-Linear Shoulder and Tail Model*: Adds both a shoulder phase at the beginning and a tailing effect at the end, capturing both delayed response and resistance in subpopulations

In essence, each model provides an additional layer of complexity to better represent nonlinear behaviors observed in microbial inactivation (*see* **Note** 7).

**Model Accuracy Criteria**

In evaluating kinetic models for bacterial inactivation, accuracy criteria are essential to determine how well a model predicts the actual inactivation behavior under specified conditions. Key criteria include statistical measures that compare predicted bacterial counts with observed values. Here is an overview of commonly used accuracy metrics:

1. Coefficient of Determination ($R^2$)

   $R^2$ measures how well the variation in observed data is explained by the model. It ranges from 0 to 1, with values closer to 1 indicating that the model accounts for most of the variability in the data. Higher $R^2$ values suggest a better fit, meaning the model predictions are closer to the actual observations:

$$R^2 = \frac{SS_R}{SS_T} = \frac{\sum_i (f_i - \overline{y})^2}{\sum_i (y_i - \overline{y})^2} \tag{8}$$

where $f_i$, $y_i$, and $\overline{y}$ are predicted data, observed data, and the average, respectively.

2. Root mean square error (RMSE)

   RMSE is the square root of the average of the squared differences between observed and predicted values. It quantifies the absolute error, providing insight into how closely the predicted bacterial counts align with the observed counts. A lower RMSE value indicates a more accurate model, as it suggests smaller deviations between the model predictions and the actual values:

$$\mathrm{RMSE} = \sqrt{\frac{1}{n}\sum_{i=1}^{n}\left(y_i - f_i\right)^2} \tag{9}$$

where $n$ is the number of observed data.

3. $A_f$ (Accuracy Factor)

   The accuracy factor ($A_f$) measures the average deviation between observed and predicted values in terms of a multiplicative factor. It is calculated as the geometric mean of the ratio between observed and predicted values. An $A_f$ value of 1 indicates a perfect match, while values greater than 1 reflect increasing levels of deviation. $A_f$ is especially useful for comparing model accuracy across different datasets and is often used in food safety modeling:

$$A_f = 10^{\left(\frac{1}{n}\sum_{i=1}^{n}\left|\log\frac{y_i}{f_i}\right|\right)} \tag{10}$$

**$B_f$ (Bias Factor)**

The bias factor ($B_f$) indicates whether the model systematically overestimates or underestimates the observed values. A $B_f$ value of 1 means the model predictions are unbiased, with neither a tendency to overpredict nor underpredict. Values less than 1 indicate a tendency toward underprediction, while values greater than 1 suggest overprediction. $B_f$ is helpful for assessing directional accuracy, ensuring the model provides realistic predictions:

$$B_f = 10^{\left(\frac{1}{n}\sum_{i=1}^{n}\left(\log\frac{y_i}{f_i}\right)\right)} \tag{11}$$

These criteria are commonly used together to assess the accuracy, reliability, and potential biases of kinetic models in bacterial inactivation studies, helping scientists and engineers select the best model for food safety and microbial control applications [7] (*see* **Note 6**).

Table 1 shows some studies in which log-linear models have been used to study the kinetics of microbial inactivation. As

**Table 1**
**Log-linear models used to study the kinetics of microbial inactivation**

| Microorganism | Media | Treatment | Condition | Kinetic model | Parameter | Model accuracy criteria | Ref |
|---|---|---|---|---|---|---|---|
| *Bacillus subtilis* spores | Sodium chloride solution | Ohmic heating | 20 V/cm, 60 kHz | LL<br>LLS | $k_{max}$: 0.79<br>S: 4.10<br>$K_{max}$:1.07 | $R^2$: 0.79<br>$R^2$: 0.80 | [8] |
| *E. coli* | Apple juice | UV-C + Heat | 5.20 J/mL – 54 °C | LS | 5D: 4.14<br>S: 0.90<br>$K_{max}$: 1.42 | $R^2$: 0.98<br>RMSE:0.0098 | [9] |
| *Listeria monocytogenes* | Milk | Thermal processing | 65 °C/1 min | LLS | D: 0.65<br>Z:11.39 | $R^2$: 0.97<br>RMSE: 0.38 | [10] |
| *S. Typhimurium* | PBS | Nonthermal plasma | 85 kV/140 s | LS | $K_{max}$: 0.20<br>D:78.84 | $R^2$: 0.99<br>RMSE: 0.16 | [11] |
| *B. cereus* spores | Citrate–phosphate buffers | Thermal processing | 100 °C/14 min | LS | D: 1.33<br>S: 0.19 | $R^2$:0.98<br>RMSE: 0.21 | [12] |
| *E. coli* | Maximum recovery diluent | Thermal processing | 60 °C/350 s | LL<br>LLS<br>LST | D: 75.2<br>D: 92.7<br>S: 10.3<br>D: 52.2<br>$N_{res}$: 4.3 | $R^2$:0.94<br>RMSE: 0.23<br>$R^2$:0.97<br>RMSE: 0.18<br>$R^2$:0.95<br>RMSE: 0.22 | [13] |
| *S. Typhimurium* | Poultry feed | Thermal processing | 95 °C/180 s | LST | $K_{max}$: 0.44<br>Nres:0.66 | $R^2$:0.99<br>RMSE:0.29 | [14] |
| Aerobic mesophilic | White cabbage | Pulsed light | 12.22 J/cm$^2$ | LL<br>LLT | $K_{max}$: 0.35<br>$K_{max}$:0.75 | $R^2$:0.80<br>RMSE: 0.37<br>$R^2$:0.87<br>RMSE:0.33 | [15] |
| *Enterococcus faecium* | Poultry feed | Thermal processing | 95 °C/180 s | LL | D:3.7 | $R^2$:0.92<br>RMSE:0.78 | [16] |
| Mold and Yeast | Cherry tomato | Sonication | 60 kHz/25 min | LL | K: 0.10<br>S: 12.10 | $R^2$:0.95<br>RMSE:0.12 | [5] |

*LL* log-linear model, *LS* log shoulder model, *LLS* log-linear + shoulder model, *LST* Log shoulder + tail model, *LLT* log-linear + tail model

mentioned in the table, $R^2$ and RMSE are the most important model accuracy criteria in the studies that have been conducted.

## 2 Materials

Software Microsoft Excel v. 2511.

## 3 Methods

To calculate the *D-value*

1. *Organize the data in Excel (three columns)*
   - Column A: Time
   - Column B: Cell count (N)
   - Column C: log10(N)
2. *Run the linear regression (log10(N) vs. Time)*
   - Go to Data → Data Analysis.
   - Select Regression.
   - In the Regression window:
     - For Input Y Range, select the log10(N) data (Column C), excluding the value corresponding to time zero.
     - For Input X Range, select the Time data (Column A).
   - Choose the output option New Worksheet Ply and then click OK.
   - Excel will automatically compute the regression. The output sheet will include a section titled "SUMMARY OUTPUT".
   - Review the following results:
     - $R^2$
     - The Intercept coefficient
     - The X Variable 1 coefficient (the slope)
   - Compute D-value using:
     - $D = -1$ / (X Variable 1)
     - In Excel: =(−1/<cell_with_X_Variable_1>)
   - The result will be in minutes or seconds, depending on the time unit used in the experiment.
3. *Create the graph*
   - Select Columns A and B.
   - Go to Insert → Chart.

- Choose Scatter (XY), then Subtype 1 (markers only, no line) (commonly used for semi-log presentation).
- Right-click the chart → Move Chart → select New sheet → OK.
- In Chart Tools (top menu), format the chart:
  - Add a title.
  - Add X- and Y-axis labels.
  - Add minor gridlines for the Y-axis.
- To set the Y-axis to a logarithmic scale:
  - Right-click the Y-axis → Format Axis.
  - Go to Axis Options/Scale → check Logarithmic scale.
- In a new chart using all points, use Excel's built-in feature to add a trendline:
  - The trendline should pass through the same points used in the regression.
  - Select the data points → right-click → Add Trendline → choose Exponential.

    Note: If one needs to ignore some points, the trendline should be created manually by drawing a straight line on the graph. To do this:
- Go to Insert → Shapes → Line.
- Draw a line spanning the *D-value* range (from time zero to *D* in minutes), and raise the left end by one log cycle.
- After defining the slope, right-click the line → Size and Properties.
- Under Size, check Lock aspect ratio and increase the height to 300%.
- Adjust the line so it aligns with the selected points.

4. *Report the equation and key parameters*
   - We will obtain the line equation and the values of $R^2$ and $D$.
   - Write the equation as:
     - $\log(C)$ = slope + intercept
   - The slope corresponds to X Variable 1, and the intercept corresponds to Intercept.
   - Place this equation just below the SUMMARY OUTPUT section (*see* **Note 8**).

To calculate the Z-value

- Open a new worksheet and enter the temperatures and their corresponding *D-values* (in minutes).
- Calculate log10($D$).
- Then, repeat the same regression procedure described in **step 2** (using log10($D$) as Y and Temperature as X).

## 4 Notes

1. Log-linear models are essential tools for designing safe and effective microbial inactivation processes in food processing or other hygienic processes.
2. The parameters of the log-linear kinetic model can disclose the thermal tolerance property of a microorganism, such that the higher *D*-value indicates higher thermal tolerance of the microorganism.
3. Since thermal processes must be designed according to the type of food (in particular, the indicator microorganism), log-linear model parameters can be used in thermal process optimization.
4. Log-linear kinetic model assumes homogeneity in microbial populations (the existence of the same thermal resistance between the microorganisms) and uniform treatment conditions (isothermal conditions), which may not always reflect real-world complexities. These two requirements are prerequisites (assumptions) for using this kinetic model.
5. Proper sample preparation and experimental setup are critical for obtaining reliable kinetic data. Differences in experimental designs, such as sample medium, temperature control, and heating methods, influence model accuracy and should be standardized.
6. Accuracy metrics like $R^2$, RMSE, $A_f$, and $B_f$ provide critical insights into the predictive performance of kinetic models. Combining these metrics ensures a comprehensive evaluation of model reliability.
7. Explore advanced models that integrate non-log-linear behaviors, such as microbial stress adaptation and recovery during thermal treatment, and heterogeneity and environmental factors.
8. Develop hybrid models that incorporate multiple inactivation factors (e.g., pH, water activity) for more accurate predictions.

## References


1. Mañas P, Pagán R (2005) Microbial inactivation by new technologies of food preservation. J Appl Microbiol 98:1387–1399. https://doi.org/10.1111/j.1365-2672.2005.02561.x
2. Peleg M (2024) Microbial inactivation kinetics models, survival curves shapes, and the temporal distributions of the individual germs deactivation. Food Eng Rev 16:163–178. https://doi.org/10.1007/s12393-024-09367-5
3. Smelt JPPM, Brul S (2014) Thermal inactivation of microorganisms. Crit Rev Food Sci Nutr 54:1371–1385. https://doi.org/10.1080/10408398.2011.637645
4. Li R, Kou X, Zhang L, Wang S (2018) Inactivation kinetics of food-borne pathogens subjected to thermal treatments: a review. Int J Hyperth 34:177–188. https://doi.org/10.1080/02656736.2017.1372643
5. Taiye Mustapha A, Zhou C, Amanor-Atiemoh-R, Owusu-Fordjour M, Wahia H, Abiola Fakayode O et al (2020) Kinetic modeling of inactivation of natural microbiota and Escherichia coli on cherry tomato treated with fixed multi-frequency sonication. Ultrason Sonochem 64:105035. https://doi.org/10.1016/j.ultsonch.2020.105035
6. Bevilacqua A, Speranza B, Sinigaglia M, Corbo M (2015) A focus on the death kinetics in predictive microbiology: benefits and limits of the Most important models and some tools dealing with their application in foods. Foods 4:565–580. https://doi.org/10.3390/foods4040565
7. Mataragas M, Drosinos EH, Vaidanis A, Metaxopoulos I (2006) Development of a predictive model for spoilage of cooked cured meat products and its validation under constant and dynamic temperature storage conditions. J Food Sci:71. https://doi.org/10.1111/j.1750-3841.2006.00058.x
8. Murashita S, Kawamura S, Koseki S (2017) Effects of Ohmic heating, including electric field intensity and frequency, on thermal inactivation of Bacillus subtilis spores. J Food Prot 80:164–168. https://doi.org/10.4315/0362-028X.JFP-16-300
9. Gouma M, Álvarez I, Condón S, Gayán E (2015) Modelling microbial inactivation kinetics of combined UV-H treatments in apple juice. Innovative Food Sci Emerg Technol 27:111–120. https://doi.org/10.1016/j.ifset.2014.11.004
10. Wang X, Zheng J, Luo L, Hong Y, Li X, Zhu Y et al (2024) Thermal inactivation kinetics of listeria monocytogenes in milk under isothermal and dynamic conditions. Food Res Int 179:114010. https://doi.org/10.1016/j.foodres.2024.114010
11. Huang M, Zhuang H, Wang J, Yan W, Zhao J, Zhang J (2018) Inactivation kinetics of salmonella typhimurium and Staphylococcus aureus in different media by dielectric barrier discharge non-thermal plasma. Appl Sci 8:2087. https://doi.org/10.3390/app8112087
12. Ruiz V, Alonso R, Salvador M, Condón S, Condón-Abanto S (2021) Impact of shoulders on the calculus of heat sterilization treatments with different bacterial spores. Food Microbiol 94:103663. https://doi.org/10.1016/j.fm.2020.103663
13. Scanlon KA, Tiwari U, Cagney C, Walsh D, McDowell DA, Duffy G (2015) Modelling the thermal inactivation of five Campylobacteraceae species. Food Control 47:135–140. https://doi.org/10.1016/j.foodcont.2014.06.042
14. Boltz TP, Moritz JS, Ayres VE, Showman CL, Jaczynski J, Shen C (2021) Modeling thermal inactivation of Salmonella Typhimurium in mash broiler feed. J Appl Poult Res 30:100208. https://doi.org/10.1016/j.japr.2021.100208
15. Izquier A, Gómez-López VM (2011) Modeling the pulsed light inactivation of microorganisms naturally occurring on vegetable substrates. Food Microbiol 28:1170–1174. https://doi.org/10.1016/j.fm.2011.03.010
16. Coe C, Boltz T, Stearns R, Foster P, Taylor RL, Moritz J et al (2022) Thermal inactivation of salmonella Typhimurium and surrogate enterococcus faecium in mash broiler feed in a laboratory scale circulated thermal bath. Poult Sci 101:101976. https://doi.org/10.1016/j.psj.2022.101976

# Chapter 10

# Estimating Inactivation Kinetic Parameters of Microorganisms Non-Log-Linear Kinetics

**Matheus Péricles Silva Láscaris and Mahmoud Yolmeh**


## Abstract

Accurate modeling of microbial inactivation is essential for advancing microbial risk assessment, food safety, and sterilization processes. Traditional log-linear models, which are based on constant inactivation rates, are unable to capture the complexities of real-world conditions, including microbial heterogeneity and dynamic environments. This chapter explores non-log-linear models, such as Weibull, biphasic, and Gompertz models, which account for variable inactivation rates, resistance variability, and phenomena such as shoulders and tails in survival curves. By providing greater predictive accuracy and flexibility, these models more accurately represent microbial responses to different stressors. Practical applications and mathematical formulations are discussed to guide accurate kinetic modeling, in addition to the use of software (GiNaFiT v. 1.6) for non-log-linear model fitting.


**Key words** Microbial inactivation, Non-log-linear model, Non-isothermal condition, Model accuracy

## 1 Introduction

The ability to accurately model microbial inactivation is a foundation in various industries' microbial risk assessment, food safety, and sterilization processes to achieve a more realistic estimate of microbial survival. For this purpose, traditional approaches have predominantly relied on log-linear models, which assume a constant inactivation rate upon exposure to the antimicrobial treatment over time [1]. However, real scenarios often deviate from these assumptions, as microbial populations exhibit diverse survival patterns influenced by environmental conditions and natural population heterogeneity. Non-log-linear microbial inactivation models have emerged as a more robust alternative, addressing these complexities and providing enhanced predictive accuracy compared to the log-linear type (*see* **Note 1**) [2]. Non-log-linear models are specifically designed to cover variations in microbial inactivation rates, a sign of processes close to reality where conditions are

Anderson S. Sant'Ana (ed.), *Heat Resistant Microorganisms in Food Safety and Spoilage*, Methods and Protocols in Food Science, https://doi.org/10.1007/978-1-0716-5280-0_10,

dynamic and populations are heterogeneous. Unlike log-linear models, these models accommodate phenomena such as biphasic survival, shoulder effects, and tailing behaviors, which are frequently observed in microbial inactivation curves. These deviations often stem from factors like microbial resistance variability, protective environmental conditions, or uneven exposure to lethal agents. By addressing these nuances, non-log-linear models align more closely with experimental observations, thereby improving their utility in practical applications [3].

The limitations of log-linear models emphasize the need for non-log-linear approaches. While log-linear models are simple and computationally efficient, their assumption of homogeneity and isothermal conditions fail to capture the complexity of microbial inactivation dynamics in systems involving diverse stressors such as heat, pressure, and chemical treatments. For instance, the survival of subpopulations with varying resistances can result in nonlinear inactivation curves, necessitating models that can describe such complex behavior. Non-log-linear models such as the Weibull, biphasic, and Gompertz models have been developed to address these challenges, offering more accurate predictions and greater flexibility [4].

This chapter discusses the evolution, importance, and practical applications of common non-log-linear models, including Weibull, biphasic, and Gompertz models, for microbial inactivation kinetic modeling. Exploring the fundamental differences between log-linear and non-log-linear approaches highlights how these advanced models enhance our understanding of microbial behavior under stress conditions. In addition, it discusses the mathematical formulations of the models and their ability to predict microbial survival in complex environments. Furthermore, this chapter studies the literature on Weibull, biphasic, and Gompertz models recently published in various food process fields.

**Definition of Non-Log-Linear Kinetic Models**

Non-log-linear kinetic models describe the behavior of microbial populations under stress conditions when inactivation rates deviate from the constant rate assumption of log-linear models. Key characteristics of non-log-linear microbial inactivation kinetic models are as follows [3]:

- *Variable inactivation rates*: Unlike log-linear kinetics, non-log-linear models accommodate changing inactivation rates over time, which are more reflective of real conditions.

- *Population heterogeneity*: Microorganisms exhibit varying resistance to stress, necessitating models that address biphasic or tailing behavior.
- *Enhanced predictive accuracy*: These models often yield better fits to experimental data, providing more accurate predictions of microbial survival.

## Log-Linear vs Non-Log-Linear Kinetic Models

In microbiology, modeling microbial inactivation is essential for predicting how populations of microorganisms respond to various stressors like heat, pressure, or chemical treatments. The two primary approaches to describe these dynamics are log-linear and non-log-linear kinetic models, each with distinct characteristics and applications (Table 1). Understanding their differences helps researchers to choose the right model for specific processes [4].

### Limitations of Log-Linear Model

The log-linear model is a fundamental approach in microbial inactivation kinetics that assumes a constant rate of microbial death over time when exposed to a lethal agent, such as heat, pressure, or a chemical. This model represents the relationship between microbial population reduction and time as a straight line on a semi-logarithmic scale. Its simplicity and ease of application make it a widely used model in food safety, sterilization, and pharmaceutical processes. Key characteristics of the log-linear model include its reliance on first-order kinetics and its assumption of uniform microbial susceptibility, meaning all microorganisms in a population are equally sensitive to the treatment and isothermal condition of treatment (uniform treatment conditions [5]. Despite its advantages, the log-linear model has notable limitations. One major drawback is its inability to accurately predict microbial survival behavior when deviations from strict linearity occur, such as tailing or shoulder effects. For instance, some microbial populations exhibit initial resistance (shoulder) before significant inactivation begins, or they may leave a resistant subpopulation that declines

**Table 1**
**Key differences between log-linear and non-log-linear kinetic models**

| Aspect | Log-linear model | Non-log-linear models |
|---|---|---|
| Inactivation rate | Constant | Variable (time-dependent) |
| Population homogeneity | Assumes uniform sensitivity | Accounts for heterogeneity |
| Curve shape | Straight line on a semi-log plot | Nonlinear (e.g., sigmoidal, biphasic) |
| Fit to experimental data | Poor fit for data with shoulders/tails | Better fit for real-world microbial data |
| Applicability | Simple, ideal conditions | Complex, dynamic conditions |

**Table 2**
**Common non-log-linear models to use for microbial kinetic inactivation modeling**

| Model | Equation | Parameter | Characteristic |
|---|---|---|---|
| Weibull | $\ln\left(\frac{N_t}{N_0}\right) = -\left(\frac{t}{\delta}\right)^{\rho}$ | Variable (time-dependent) | Covering different shapes of inactivation curves by introducing a shape parameter<br>Ability to describe both shoulders and tails<br>Suitable for complex heat inactivation patterns |
| Biphasic | $\ln\left(\frac{N_t}{N_0}\right) = \ln((1-f)\exp(k_1 t) + f\ \exp(-k_2 t))$ | Nonlinear (e.g., sigmoidal, biphasic) | Division of a microbial population into two or more subpopulations with different inactivation rates<br>Suitable for microbial populations with different thermal resistances |
| Gompertz | $S(t) = \exp\left[-e^{\alpha - bt}\right]$ | Better fit for real-world microbial data | Flexible models and suitable for sigmoidal (S-shaped) inactivation curves |

more slowly, causing tailing. These deviations often arise due to factors such as microbial heterogeneity, protective environmental conditions, or uneven distribution of the lethal agent [6]. Consequently, the log-linear model may oversimplify microbial inactivation dynamics, leading to less precise predictions and potentially compromising the safety margins in critical processes. Therefore, it is important to use kinetic models that can accurately predict the non-log-linear survival behaviors of microorganisms (Table 2).

### Non-Log-Linear Kinetic Models

Based on the literature review, the Weibull, biphasic, and Gompertz models are the common non-log-linear models used to study microbial inactivation kinetics. Therefore, these three models are discussed here separately.

### Weibull Model

The Weibull model has emerged as a versatile and widely used mathematical framework in microbial inactivation kinetics. It is based on the Weibull distribution, a statistical model originally used to describe the survival patterns of microorganisms when subjected to various stress factors such as heat, pressure, or chemical treatments. Unlike traditional first-order kinetics, the Weibull model captures the nonlinear and often complex survival curves that exist in real applications. In microbial inactivation, the Weibull model covers the variability in microorganisms' resistance to inactivation processes [7].

### Equation and Its Derivation

The Weibull model is based on the Weibull distribution, expressed as follows:

$$f(t) = \frac{\rho}{\beta}\left(\frac{t}{\beta}\right)^{\rho-1} e^{-\left(\frac{t}{\beta}\right)^{\rho}}$$

where $f(t)$, $t$, $\rho$, and $\beta$ represent the probability density functions describing the likelihood of inactivation at time t, time, shape parameter, and scale parameter, respectively.

The cumulative distribution function gives the proportion of the population inactivated by time t:

$$F(t) = 1 - e^{-\left(\frac{t}{\beta}\right)^{\rho}}$$

Here, $F(t)$ represents the fraction of the microbial population inactivated by time $t$.

The survival function $S(t)$, which describes the fraction of the microbial population still alive (surviving) at time $t$, is derived as

$$S(t) = 1 - F(t)$$

Substituting $F(t)$, the previous equation:

$$S(t) = 1 - \left(1 - e^{-\left(\frac{t}{\beta}\right)^{\rho}}\right)$$

$$S(t) = e^{-\left(\frac{t}{\beta}\right)^{\rho}}$$

or:

$$\text{loglog } N_t = \text{loglog } N_0 - \left(\frac{t}{\beta}\right)^{\rho}$$

where $N_0$, $N$, $t$, $\beta$, and $\rho$ are the initial microbial population, microbial population after $t$, time, the scale parameter related to the time required for a significant reduction, and the shape parameter, which determines the curve's shape (concave or convex) [5].

The different forms of the curve's shape are as follows:

$\rho$ =1: The model reduces to a linear inactivation model (log-linear) (*see* **Note 2**).

$\rho > 1$: The curve is convex, indicating faster initial inactivation.

$\rho < 1$: The curve is concave, reflecting slower initial inactivation (*see* **Note 3**).

Since this model cannot adequately describe survival curves with a "tail," where some microbes persist at low levels for extended periods, there is the developed version of this model entitled "Weibull with tail." (*see* **Note 4**) This is an extension of the standard Weibull model designed to account for the tailing behavior, where a subpopulation of microbes demonstrates higher resistance and survives for a prolonged period. The survival curve typically flattens at the end, showing a residual population that is difficult to inactivate. This extended version is useful for describing microbial inactivation in complex systems where some cells enter a dormant state, form spores, or are shielded from the inactivation process.

$$\text{loglog } N_t = \text{loglog } (N_0 - N_r) - \left(\frac{t}{\beta}\right)^{\rho} + N_r$$

where $N_r$ represents the residual population that remains unaffected by the treatment [7].

**Key Features**

- The model acknowledges the heterogeneity within the microbial population, reflecting the coexistence of sensitive and resistant subpopulations.
- It captures the flattening of the survival curve, improving the accuracy of predictions for processes where tailing is observed.

**Applications**

- Useful for describing microbial inactivation in complex systems where some cells enter a dormant state, form spores, or are shielded from the inactivation process.

**Biphasic Model**

The biphasic model describes microbial inactivation when two subpopulations with differing resistances to stress coexist (*see* **Note 6**). It provides a more accurate depiction of survival curves that deviate from simple first-order kinetics, typically showing two phases of inactivation: a fast initial phase followed by a slower phase [8].

**Concept of the Biphasic Model**

Microbial populations are often heterogeneous, containing a mixture of:

1. *Sensitive subpopulation*: Quickly inactivated under stress
2. *Resistant subpopulation*: More robust, surviving longer under the same conditions

The biphasic model assumes that the survival curve is the sum of these two distinct first-order decay processes.

- *Survival of individual subpopulations*: The survival of each subpopulation follows first-order kinetics, described by:

$$S_1(t) = f_1 e^{-k_1 t}$$

$$S_2(t) = f_2 e^{-k_2 t}$$

where

- $S_1(t)$: Fraction of the sensitive subpopulation surviving at time $t$,
- $S_2(t)$: Fraction of the resistant subpopulation surviving at time $t$,
- $f_1$: Initial fraction of the sensitive subpopulation ($f_1 + f_2 = 1$),
- $f_2$: Initial fraction of the resistant subpopulation,
- $k_1$: Inactivation rate constant for the sensitive subpopulation ($k_1 > k_2$),
- $k_2$: Inactivation rate constant for the resistant subpopulation.
- *Overall survival function*: The total survival fraction $S(t)$ at any time $t$ is the weighted sum of the survival functions for the two subpopulations:

$$S(t) = S_1(t) + S_2(t)$$

Substituting the survival expressions for $S_1(t)$ and $S_2(t)$:

$$S(t) = f_1 e^{-k_1 t} + f_2 e^{-k_2 t}$$

This is the general form of the biphasic model.

- *Special cases*
  - If $f_2 = 0$: The population is entirely sensitive, and the model reduces to a single-phase first-order inactivation:

$$S(t) = e^{-k_1 t}.$$

  - If $f_1 = 0$: The population is entirely resistant, and the model reduces to:

$$S(t) = e^{-k_2 t}.$$

  - If $k_1 \approx k_2$: The curve approximates a single-phase decay, but this is rarely observed in true biphasic kinetics.
- *Logarithmic transformation*: For data fitting, the equation can be expressed in logarithmic form:

$$\ln\ S(t) = \ln\ln\left(f_1 e^{-k_1 t} + f_2 e^{-k_2 t}\right)$$

This transformation is useful for evaluating the rate constants and population fractions using experimental data.

- *Interpretation of parameters*
  - $f_1$ and $f_2$: Indicate the proportion of sensitive and resistant populations, respectively.
  - $k_1$ and $k_2$: Provide insights into the relative susceptibility of each subpopulation to inactivation treatments (*see* **Note 5**) [5].

**Gompertz Model (S-Shaped Curve)**

The Gompertz model, originally developed to describe population growth, has been adapted to microbial deactivation studies (*see* **Note 7**). This sigmoidal model captures the nonlinear progression of microbial death, beginning with a lag phase where deactivation is minimal, followed by an accelerated phase, and finally leveling off as the population nears complete deactivation. Temporal patterns from the Gompertz model are particularly valuable for predicting outcomes in systems where initial resistance or adaptation delays the deactivation process [9].

The Gompertz model is widely used in microbial inactivation kinetics to describe nonlinear, S-shaped survival curves. It is particularly effective in modeling processes where microbial populations exhibit an initial lag phase, followed by accelerated inactivation, and then a decelerated phase as the treatment progresses [10].

**Concept of the Gompertz Model**

The Gompertz model is derived from a growth curve model that characterizes processes where the rate of change depends on the remaining population and time. For microbial inactivation, it adapts to reflect population decline, resulting in an S-shaped inactivation curve. There are at least three phases in the concept of this model including I) a shoulder (no decline in cell population), II) a death phase as exponential (known by the inactivation rate), and III) a tail, i.e., a residual population [5].

Before addressing the derivation of the equation of the Gompertz model in microbial inactivation kinetic modeling, the main reasons for the formation of the shoulder and tail in the microbial death curve upon exposure to a lethal agent are explained. There are several potential reasons for the presence of a shoulder in the death curve of a microorganism, including

- The protective effect of the food or medium components (fats, proteins) on cells.
- Creating a clump structure and placing cells in it.

- Cumulative damage required before cell inactivation.
- It can be related to the adaptation potential of the microorganism, which microorganism is trying to synthesize vital components during this phase (in contrast to the destruction of vital components, their synthesis also exists).

On the other hand, the presence of a tail at the final stage of inactivation kinetics implies the existence of a sub-resistant population. This phase can be described according to the presence of a high-resistant sub-population to heat/inactivation treatment, or it could be related to the unavailability of this group to the lethal treatment or the treatment not in sufficient dose for this group [5].

- General form of the Gompertz function: The Gompertz model is generally expressed as:

$$S(t) = \exp\left[-e^{\alpha - bt}\right]$$

where $S(t)$, $\alpha$, $b$, and $t$ represent the fraction of the surviving microbial population at time t, a constant related to the upper asymptote (initial survival state) and lag phase, constant rate determining the curve slope, and time, respectively.

- *Starting with a logistic process*: The Gompertz function can be derived as a special case of logistic-like behavior, where the survival rate is proportional to both the population size and the logarithm of the population.

**Differential Equation Representation**

The survival curve can be described by a nonlinear differential equation:

$$\frac{d \ln S(t)}{dt} = -be^{\alpha - bt}$$

where $\frac{d \ln S(t)}{dt}$ is the instantaneous rate of survival decline and $-be^{a - bt}$ introduces nonlinearity, reflecting the accelerated and decelerated inactivation phases.

- *Integrating the differential equation*: To find the cumulative survival function $S(t)$, integrate the above equation:

$$\ln S(t) = -e^{\alpha - bt} + C$$

The constant $C$ is the initial condition for $S(0) = 1$, based on ln $S(0) = 0$; therefore, $C = 0$.

Thus:

$$\ln S(t) = -e^{\alpha - bt}$$

then, considering the exponential of both sides:

$$S(t) = \exp\left[-e^{\alpha - bt}\right]$$

### Interpretation of Parameters

$\alpha$: Related to the lag time. Larger $\alpha$ indicates a longer lag phase where microbial inactivation begins slowly.

$b$: Determines the steepness of the survival curve, controlling how quickly the population moves through the accelerated inactivation phase.

$t$: Time variable, representing the duration of the inactivation process.

### Key Characteristics of the Gompertz Model

1. *S-shaped curve*
   - Initial lag phase with minimal inactivation
   - Rapid inactivation phase where most of the population is eliminated
   - Plateau phase where further inactivation slows significantly
2. *Flexibility*
   - Captures complex survival behavior better than simple first-order models
   - Can be adjusted for various treatments, including thermal and nonthermal processes [9]

### Special Cases and Variations

1. *Linear transformation*: The Gompertz model is often transformed to simplify parameter estimation:

$$\ln\ln\left[S(t)\right] = \alpha - bt$$

This linearized form facilitates the regression analysis of experimental data.

2. *Modified Gompertz model*: Extensions of the model include additional parameters to capture specific features, such as tailing or multi-phase inactivation [5].

Table 3 presents several studies conducted in recent years that employed non-log-linear models to predict microbial inactivation. As shown in the table, *Escherichia coli*, *Listeria monocytogenes*, and *Staphylococcus aureus* are the primary pathogenic bacteria investigated. This focus is likely due to the significant pathogenicity of these bacteria and/or their high potential for transmission through food. $R^2$ and RMSE are the most commonly used criteria for

**Table 3**
**Some studies performed to predict microbial inactivation through non-log-linear models in recent years**

| Microorganism | Media | Treatment | Method | Kinetic model | Parameter | Model accuracy criteria | Concluding remark | Ref |
|---|---|---|---|---|---|---|---|---|
| *L. monocytogenes*<br>*E. coli* O157:H7 | Soybean sprouts | Chemical sanitizers with ultrasound | HCl (5%) + NaCl (2 M) + fumaric acid (0.5%) + US (40 kHz/ 23 °C)—15 min | WM | p: 0.25<br>δ: 0.12<br>p: 0.27<br>δ: 0.19 | $R^2$: 0.97<br>RMSE: 0.31<br>$R^2$: 0.94<br>RMSE: 0.36 | In this study, the WM model fit the data appropriately, showing the exact contact time required to inactivate the pathogens. | [17] |
| *L. monocytogenes* | Stainless steel coupons | Neutral electrolyzed | Available chlorine concentration 200 mg/ mL—360 s | WM | p: 0.37<br>δ: 0.91 | $R^2$: 0.96<br>RMSE:0.11 | WM indicates the optimization of the use of the compound as an eco-friendly sanitizer to be used in the food industry. | [18] |
| *S. aureus* | GTYE broth | Low thermal processing | 61 °C/180 min | WM | p: 0.86<br>δ: 6.85 | $R^2$: 0.98<br>RMSE: 0.27 | WM proved effective in predicting inactivation at moderate temperatures. | [19] |
| *L. monocytogenes*<br>*S. aureus* | Citrate-phosphate McIlvaine buffer | Pulsed electric field | 35 kV/cm (pH 3.5)—600μs | WM | p: 0.36<br>δ: 21.39<br>p: 0.11<br>δ: 0.11 | $R^2$: 0.93<br>RMSE: 0.16<br>$R^2$: 0.99<br>RMSE: 0.18 | The model indicates that the treatment alone does not exhibit lethal efficacy, and there is a need to define additional hurdles to enhance | [20] |

(continued)

**Table 3 (continued)**

| Microorganism | Media | Treatment | Method | Kinetic model | Parameter | Model accuracy criteria | Concluding remark | Ref |
|---|---|---|---|---|---|---|---|---|
| | | | | | | | lethality against *L. monocytogenes*. | |
| *S. enterica*<br><br>*B. cereus* | Instant soup | γ-Irradiation | 10 kGy | WM | p: 0.74<br>δ: 1.29<br>p: 2.37<br>δ: 0.5 | $R^2$: 0.97<br><br>$R^2$: 0.98 | The model demonstrated a better fit to the data compared to other tested models, indicating that irradiation is effective for pathogen elimination. | [21] |
| *S. aureus* | Ballast water | UV-C | 400 mJ/cm$^2$ | BM | $k_1$: 0.15<br>$k_2$: 0.00 | $R^2$: 0.98<br>RMSE: 0.26 | The BM showed a more suitable fit for ultraviolet disinfection compared to the other three models, indicating a more sensitive phase of the microorganism to treatment (exponential phase > stationary phase > lag phase). | [22] |
| *E. coli* O157:H7 | Ceberg lettuce | Electrochemical disinfection | 216 mA/cm$^2$—600 mg/L—120 min | BM | $k_1$: 0.55<br>$k_2$: 0.06 | $R^2$: 0.97<br>RMSE: 0.37 | The model proved suitable for demonstrating that | [23] |

| | | | | | | | | |
|---|---|---|---|---|---|---|---|---|
| | | | | | | | this technology is effective in maintaining the quality of wash water. | |
| *B. amyloliquefaciens* | Deionized water<br>Green pea purée | Pressure-assisted thermal processing | 700 Mpa/ 121 °C—20 min | BM | $k_1$:16.26<br>$k_2$: 0.54<br>$k_1$:5.84<br>$k_2$:0.13 | $R^2$:0.98<br>RMSE:0.1<br>$R^2$:0.99<br>RMSE:0.03 | Due to the inactivation behavior of this technology—Characterized by rapid inactivation due to pressure in the initial phase, followed by slower inactivation by the remaining spores—The model provided more reliable indicators than other models. | [24] |
| *S. senftenberg* | Liquid egg | High Hydrostatic pressure | 200 Mpa/13 min | BM | $k_1$:0.23<br>$k_2$:0.24 | $R^2$: 0.9<br>RMSE: 0.2 | The data from the model indicate that the treatment is as effective as conventional thermal pasteurization. | [25] |
| Aspergillus section Nigri spores | Sodium hypochlorite | Chemical sanitizers | 700 mg/L (pH 6.3)/ 60 min | BM | $k_1$: 2.55<br>$k_2$: 0.51 | $R^2$: 0.99<br>RMSE: 0.01 | The model indicates a disadvantage in using this chemical | [26] |

(continued)

**Table 3 (continued)**

| Microorganism | Media | Treatment | Method | Kinetic model | Parameter | Model accuracy criteria | Concluding remark | Ref |
|---|---|---|---|---|---|---|---|---|
| | | | | | | | compound compared to sodium hypochlorite, as it requires higher concentration and longer exposure time. | |
| *L. innocua* | Vegetable beverage | Low-temperature treatments | 50 °C/75 min | GM[a] | $\mu_{max}$: 0.13 | $R^2$: 0.99 RMSE: 0.19 | GM[a] provided a better fit to the data, even at different times, compared to the other models tested. | [11] |
| | | | 55 °C/75 min | | $\mu_{max}$: 0.87 | $R^2$: 0.99 RMSE: 0.16 | | |
| | | | 60 °C/75 min | | $\mu_{max}$: 5.33 | $R^2$: 0.99 RMSE: 0.17 | | |
| | | | 65 °C/75 min | | $\mu_{max}$: 16.25 | $R^2$: 0.96 RMSE: 0.20 | | |
| | Milk | High-pressure processing | 300 MPa/10 min | GM | $\mu_{max}$: 0.09 | $R^2$: 0.89 | In this study, GM underestimated the lethality of the process; however, this limitation had no practical effect, as the microbial count was below the predictions. | [27] |
| | | | 400 MPa/10 min | | $\mu_{max}$: 0.32 | $R^2$: 0.96 | | |
| | | | 500 MPa/10 min | | $\mu_{max}$: 2.37 | $R^2$: 0.97 | | |
| | | | 600 MPa/10 min | | $\mu_{max}$: 7.64 | $R^2$: 0.98 | | |

| | | | | | | | | |
|---|---|---|---|---|---|---|---|---|
| *E. coli* | Sliced tomato | Pulsed light | 8 $J/cm^2$ | GM | $\mu_{max}$: −0.18 | $R^2$: 0.93 | Negative values of μmax show that the inactivation of *E. coli* was accelerated after the treatments, indicating the effectiveness of the treatment. | [28] |
| *S. typhimurium* | Carrot juice | High pressure | 10 Mpa/90 min | GM[a] | $\mu_{max}$: 0.19 | $R^2$: 0.99<br>RMSE: 0.16 | The study data showed a better fit to the GM model, indicating that the treatment was effective in inactivating the microorganism. | [29] |
| *E. coli O157:H7*<br>*L. monocytogenes* | Almond milk | Ultrasound | 20 kHz/8 min | GM | $\mu_{max}$: 1.19<br>$\mu_{max}$: 0.79 | $R^2$: 0.97<br>$R^2$: 0.94 | The data fitted to the model indicate that the process significantly reduced the levels of both microorganisms. | [30] |

*WM* Weibull model, $p$ shape parameter, $\delta$ scale parameter, *BM* biphasic model, $k_1$ and $k_2$ inactivation rates constants for the two distinct phases, *GM* Gompertz model, $\mu_{max}$ maximal growth rate

[a]Model modified

evaluating the accuracy of models in studies on microbial inactivation using non-log-linear models (*see* **Note 8**). Higher values of $R^2$ (closer to one) and lower values of RMSE (closer to zero) indicate greater model efficiency.

It is worth mentioning that non-log-linear models have been successfully applied to model bacterial inactivation through advanced thermal methods, such as ohmic heating [11, 12], as well as nonthermal techniques, including ultrasound [13, 14] and high hydrostatic pressures [15, 16].

## 2 Materials

Microsoft Excel software v. 2511 with the GinaFiT add-in v. 1.6 (for installation instructions, *see* **Note 9**).

## 3 Methods

### Use of the GinaFiT Add-In

1. *Insertion of experimental data*

   Experimental data were entered directly into a Microsoft Excel worksheet and organized as follows:

   - First column: Thermal treatment time (in minutes or seconds, using a consistent unit)
   - Second column: Population of surviving microorganisms, expressed as CFU/mL (or an equivalent unit)

   This data structure is essential for the add-in to correctly identify the independent (time) and dependent (microbial survival) variables.

2. *Logarithmic transformation of survival data*

   For mathematical model fitting in GinaFiT, survival data must be expressed on a logarithmic scale:

   - A new column was created in the worksheet.
   - The decimal logarithm of the surviving population was calculated using the Excel function LOG10(N).
   - This calculation was applied to all experimental data points.

3. *Data selection for modeling*

   - Both the time column and the log10-transformed survival column were selected simultaneously.

   *Note:* Only log-transformed data should be used for model fitting. The use of non-log-transformed data may result in errors or incorrect model adjustment.

4. *Model fitting using GinaFiT*
   - The Add-ins tab in Excel was accessed.
   - The GinaFiT tool was selected.
   - A window displaying the available mathematical models (e.g., log-linear, Weibull, and models with shoulder and/or tail) was opened.
   - The desired model was selected, and the fitting procedure was initiated.

   After model selection, GinaFiT automatically generated a new worksheet containing the fitting results.

5. *Analysis of the output worksheet*

   The worksheet generated by GinaFiT included:
   - $R^2$ (coefficient of determination)
   - Adjusted $R^2$
   - The mathematical equation of the selected model
   - The estimated model parameters
   - Observed and predicted values
   - A graph showing the model fit to the experimental data

   These outputs allow evaluation of both the statistical quality of the fit and the biological adequacy of the selected model.

6. *Identification of fitting errors*

   If the selected model was not compatible with the experimental data, GinaFiT automatically displayed an error message in the worksheet or as a pop-up window.

   This error indicates that the data do not meet the mathematical assumptions of the chosen model (e.g., the presence of a tail or shoulder incompatible with a simple log-linear model).

7. *Assessment of model fit quality*

   To determine whether the model adequately describes the experimental data, the following parameters were evaluated:
   - $R^2$
   - Adjusted $R^2$

8. *Testing alternative models (if necessary)*

   If the model fit was unsatisfactory or biologically inconsistent:
   - The procedure was resumed from **step 3**.
   - The data were reselected.
   - An alternative model available in GinaFiT was tested.

## 4 Notes

1. Non-log-linear models provide more accurate predictions of microbial inactivation compared to log-linear models, particularly under real-world conditions where microbial populations exhibit heterogeneity and dynamic conditions.
2. The Weibull model can be simplified to log-linear under specific conditions. This occurs when the Weibull distribution's shape parameter ($\rho$) equals 1.
3. In the Weibull model, if $\rho < 1$, the curve will be in concave form, which reflects a slower initial inactivation. However, if $\rho > 1$, the curve will be in convex form, which indicates faster initial inactivation.
4. The Weibull model is suitable for describing complex heat inactivation and tailing behaviors.
5. The difference value between $k_1$ and $k_2$ provides insight into the heterogeneity of the population's resistance. A larger difference implies a significant disparity in susceptibility between the two subpopulations. As this difference gets closer to zero, the curve will be approximately a single-phase decay.
6. The biphasic model is useful, especially for dividing microbial populations into sensitive and resistant subpopulations, each following distinct first-order decay rates.
7. The Gompertz model is useful for systems, in which initial microbial resistance or adaptation influences inactivation kinetics.
8. The use of accuracy criteria, such as $R^2$ and RMSE, is crucial to evaluate the predictive potentials in modeling microbial inactivation studies using non-log-linear models.
9. The file was sent together with the activity materials. However, it can also be obtained through the link https://cit.kuleuven.be/biotec/software/GinaFit, requiring only the entry of personal information to submit the request.

### References

1. Dolan KD, Valdramidis VP, Mishra DK (2013) Parameter estimation for dynamic microbial inactivation: which model, which precision? Food Control 29:401–408. https://doi.org/10.1016/j.foodcont.2012.05.042
2. Corradini MG, Peleg M (2009) Dynamic model of heat inactivation kinetics for bacterial adaptation. Appl Environ Microbiol 75:2590–2597. https://doi.org/10.1128/AEM.02167-08
3. Peleg M (2024) Microbial inactivation kinetics models, survival curves shapes, and the temporal distributions of the individual germs deactivation. Food Eng Rev 16:163–178. https://doi.org/10.1007/s12393-024-09367-5
4. Li R, Kou X, Zhang L, Wang S (2018) Inactivation kinetics of food-borne pathogens subjected to thermal treatments: a review. Int J Hyperth 34:177–188. https://doi.org/10.1080/02656736.2017.1372643

5. Bevilacqua A, Speranza B, Sinigaglia M, Corbo M (2015) A focus on the death kinetics in predictive microbiology: benefits and limits of the Most important models and some tools dealing with their application in foods. Foods 4:565–580. https://doi.org/10.3390/foods4040565
6. Aragao GMF, Corradini MG, Normand MD, Peleg M (2007) Evaluation of the Weibull and log normal distribution functions as survival models of Escherichia coli under isothermal and non isothermal conditions. Int J Food Microbiol 119:243–257. https://doi.org/10.1016/j.ijfoodmicro.2007.08.004
7. Buzrul S (2022) The Weibull model for microbial inactivation. Food Eng Rev 14:45–61. https://doi.org/10.1007/s12393-021-09291-y
8. Corradini MG, Normand MD, Peleg M (2007) Modeling non-isothermal heat inactivation of microorganisms having biphasic isothermal survival curves. Int J Food Microbiol 116:391–399. https://doi.org/10.1016/j.ijfoodmicro.2007.02.004
9. Gil MM, Miller FA, Brandão TRS, Silva CLM (2011) On the use of the Gompertz model to predict microbial thermal inactivation under isothermal and non-isothermal conditions. Food Eng Rev 3:17–25. https://doi.org/10.1007/s12393-010-9032-2
10. Gil MM, Miller FA, Brandão TRS, Silva CLM (2017) Mathematical models for prediction of temperature effects on kinetic parameters of microorganisms' inactivation: tools for model comparison and adequacy in data fitting. Food Bioprocess Technol 10:2208–2225. https://doi.org/10.1007/s11947-017-1989-x
11. Pereira MO, Guimarães JT, Ramos GLPA, do Prado-Silva L, Nascimento JS, Sant'Ana AS et al (2020) Inactivation kinetics of listeria monocytogenes in whey dairy beverage processed with ohmic heating. LWT 127:109420. https://doi.org/10.1016/j.lwt.2020.109420
12. Murashita S, Kawamura S, Koseki S (2017) Effects of Ohmic heating, including electric field intensity and frequency, on thermal inactivation of Bacillus subtilis spores. J Food Prot 80:164–168. https://doi.org/10.4315/0362-028X.JFP-16-300
13. Johnson Esua O, Sun D-W, Ajani CK, Cheng J-H, Keener KM (2022) Modelling of inactivation kinetics of Escherichia coli and Listeria monocytogenes on grass carp treated by combining ultrasound with plasma functionalized buffer. Ultrason Sonochem 88:106086. https://doi.org/10.1016/j.ultsonch.2022.106086
14. Alenyorege EA, Ma H, Ayim I (2019) Inactivation kinetics of inoculated *Escherichia coli* and *Listeria innocua* in fresh-cut Chinese cabbage using sweeping frequency ultrasound. J Food Saf:39. https://doi.org/10.1111/jfs.12696
15. Maresca P, Ferrari G (2017) Modeling of the microbial inactivation by high hydrostatic pressure freezing. Food Control 73:8–17. https://doi.org/10.1016/j.foodcont.2016.05.047
16. Guillou S, Membré J-M (2019) Inactivation of listeria monocytogenes, Staphylococcus aureus, and salmonella enterica under high hydrostatic pressure: a quantitative analysis of existing literature data. J Food Prot 82:1802–1814. https://doi.org/10.4315/0362-028X.JFP-19-132
17. Ngnitcho P-FK, Tango CN, Khan I, Daliri EB-M, Chellian R, Oh DH (2018) The applicability of Weibull model for the kinetics inactivation of Listeria monocytogenes and Escherichia coli O157: H7 on soybean sprouts submitted to chemical sanitizers in combination with ultrasound at mild temperatures. LWT 91:573–579. https://doi.org/10.1016/j.lwt.2018.01.073
18. Possas A, Pérez-Rodríguez F, Tarlak F, García-Gimeno RM (2021) Quantifying and modelling the inactivation of Listeria monocytogenes by electrolyzed water on food contact surfaces. J Food Eng 290:110287. https://doi.org/10.1016/j.jfoodeng.2020.110287
19. Lehotová V, Miháliková K, Medveďová A, Valík Ľ (2021) Modelling the inactivation of Staphylococcus aureus at moderate heating temperatures. Czech J Food Sci 39:42–48. https://doi.org/10.17221/201/2020-CJFS
20. Saldaña G, Puértolas E, Condón S, Álvarez I, Raso J (2010) Inactivation kinetics of pulsed electric field-resistant strains of Listeria monocytogenes and Staphylococcus aureus in media of different pH. Food Microbiol 27:550–558. https://doi.org/10.1016/j.fm.2010.01.002
21. Apaydın D, Tırpancı Sivri G, Demirci AŞ (2023) Modeling the γ-irradiation inactivation kinetics of foodborne pathogens *Escherichia coli* O157:H7, *Salmonella, Staphylococcus aureus and Bacillus cereus* in instant soup. Food Sci Technol Int. https://doi.org/10.1177/10820132231210317
22. Ren Z, Zhang L, Shi Y (2016) UV disinfection of Staphylococcus aureus in ballast water: effect of growth phase on the disinfection kinetics and the mechanization at molecular level. Sci China Technol Sci 59:330–336. https://doi.org/10.1007/s11431-015-5965-2
23. López-Gálvez F, Posada-Izquierdo GD, Selma MV, Pérez-Rodríguez F, Gobet J, Gil MI et al

(2012) Electrochemical disinfection: an efficient treatment to inactivate Escherichia coli O157:H7 in process wash water containing organic matter. Food Microbiol 30:146–156. https://doi.org/10.1016/j.fm.2011.09.010

24. Ahn J, Balasubramaniam BVM (2014) Inactivation kinetics and injury recovery of Bacillus amyloliquefaciens spores in low-acid foods during pressure-assisted thermal processing. Food Sci Biotechnol 23:1851–1857. https://doi.org/10.1007/s10068-014-0253-7

25. Ramos SJ, Sanz M, Alonso D, Virto R (2019) Estimation of the pasteurization treatment of liquid whole egg inoculated with Salmonella Senftenberg 775W by high hydrostatic pressure treatments. Int J Eng Sci 8:19–28. https://doi.org/10.9790/1813-0806011928

26. Frison L, Silvina S, Verónica F, María B (2015) Susceptibility of black aspergili conidia to industrial sanitizers. Int Res J Public Environ Health 2:65–69

27. Serment-Moreno V, Fuentes C, Torres JA, Welti-Chanes J (2017) A Gompertz model approach to microbial inactivation kinetics by high-pressure processing (HPP): model selection and experimental validation. J Food Sci 82:1885–1891. https://doi.org/10.1111/1750-3841.13783

28. Valdivia-Nájar CG, Martín-Belloso O, Giner-Seguí J, Soliva-Fortuny R (2017) Modeling the inactivation of Listeria innocua and Escherichia coli in fresh-cut tomato treated with pulsed light. Food Bioprocess Technol 10:266–274. https://doi.org/10.1007/s11947-016-1806-y

29. Liao H, Kong X, Zhang Z, Liao X, Hu X (2010) Modeling the inactivation of Salmonella typhimurium by dense phase carbon dioxide in carrot juice. Food Microbiol 27:94–100. https://doi.org/10.1016/j.fm.2009.09.012

30. Iorio MC, Bevilacqua A, Corbo MR, Campaniello D, Sinigaglia M, Altieri C (2019) A case study on the use of ultrasound for the inhibition of Escherichia coli O157:H7 and Listeria monocytogenes in almond milk. Ultrason Sonochem 52:477–483. https://doi.org/10.1016/j.ultsonch.2018.12.026

# Chapter 11

# Calibration of Thermocouples for Temperature Measurement During Thermal Processes

**Matheus S. Mourão and Roberto Pessanha da Silva Pires**


## Abstract

The calibration of thermocouples is a crucial process for achieving a process with linear and accurate characteristics. In this chapter, we discuss the principal attributes of thermocouple calibration and various methods for carrying out this process. The calibration process can present different aspects of realization. In some cases, this process is performed in a straightforward manner using basic instruments. At other times, this process is performed using software and other necessary devices. In other cases, this process is performed with specialized devices designed for this purpose. This chapter presents some of these applications and provides a step-by-step guide on how to apply them.


**Key words** Thermoelectric effect, Seebeck coefficient, Arduino

## 1 Introduction

Thermocouples are appropriate for temperature measurement and are based on the thermoelectric effect. These devices consist of two dissimilar wires at the end of a small, compact measuring junction. In this sense, [1] was the first to observe that the thermocouple created a current in a closed circuit when one junction was at a different temperature from the other (Fig. 1). During the use of these devices, a thermoelectric electromotive force (EMF) is generated and measured across the two opposite ends. However, in all devices, the EMF is created along the length of the wires, in any cables subjected to a temperature gradient [2, 3].

There are different types of thermocouples, including types J, K, E, S, and T, which have distinct characteristics in terms of composition and operating temperature (Table 1). Thus, calibration is a crucial mechanism for ensuring accurate operation. The calibration process is typically realized by decreasing the temperature from the major to the low calibration temperature. However, in some cases, if the device is subject to calibration temperature

Anderson S. Sant'Ana (ed.), *Heat Resistant Microorganisms in Food Safety and Spoilage*, Methods and Protocols in Food Science, https://doi.org/10.1007/978-1-0716-5280-0_11,

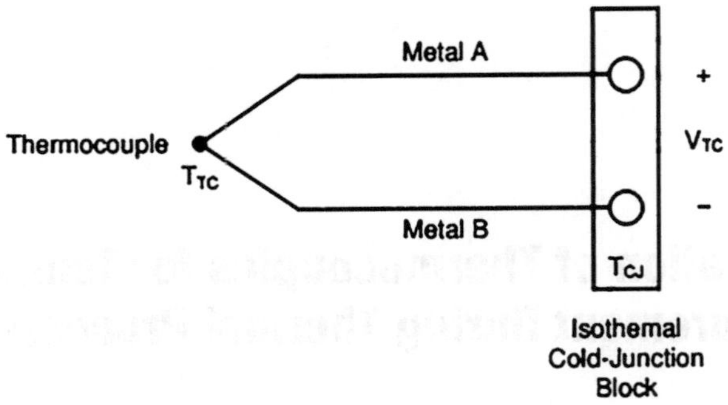


**Fig. 1** Thermocouple voltage. (Source: Wu [4])

**Table 1**
**Characteristics of thermocouple types**

| Thermocouple type | Lead metal A (+) | Lead metal B (−) | Temperature range (°C) | EMF over temperature range (mV) | Seebeck coefficient (μV/°C at 0 °C) |
|---|---|---|---|---|---|
| J | Iron | Constantan | −210 to 1200 | −8.095 to 69.553 | 50.37 |
| K | Chromel | Alumel | −270 to 1370 | −6.458 to 54.886 | 39.48 |
| T | Copper | Constantan | −200 to 400 | −6.258 to 20.872 | 38.74 |
| E | Chromel | Constantan | −270 to 1000 | −9.385 to 76.373 | 58.70 |
| S | Platinum and 10% rhodium | Platinum | −50 to 1768 | −0.236 to 18.693 | 10.19 |

Source: Wu [4]

changes in cycles of 500 °C or above, such as type K thermocouples, the calibration should be performed at increasing temperatures, and the first point can be checked and repeated at the end point [5, 6].

The calibration process is related to essential characteristics of thermocouples, like the temperature being measured, thermocouple aging, and the size of the wire. The temperature range of the application being measured is the first factor to consider in the calibration process. These characteristics are specific to each type of thermocouple calibration. The thermocouple aging is related to the degradation of this device over time, along with the ambient characteristics used. The alloy of thermocouples undergoes oxidation/corrosion over time. These processes can be present in a thermocouple, but in others, they will be controlled by the composition characteristic of the device. The size of the wire can decrease the upper temperature limit. Thermocouples of the type K have a superior limit at 1260 °C because of the 8-gauge wire in our composition. In other cases, where the device is produced with

24-gauge, the superior limit recommends a temperature of 870 °C. These limits apply only to the thermocouple tube and the wire type in the tube of convectional protection that has a closed end [7].

## 2 Materials

### 2.1 General Calibration Process

Thermocouples (different types (K or J)), a voltmeter with a minimum sensitivity of 10 μV, a thermometer with the scale from 0 to 100 °C and a minimum sensitivity of 1 °C, a kettle, a water stirrer, a water container, a refractory container, and a heat source. This method was described by [8].

### 2.2 Calibration of a Type K Thermocouple for Use in a Data Acquisition System with Arduino

A computer with the operating system MS-Windows Xp (equivalent or higher), MS-Visual C# 2010 Express or MS-Visual Studio software, Arduino Software 0022, NET Framework 4 Software, MAX6675 Arduino library code, 1 Arduino Mega 1280, one type K thermocouple; one mercury thermometer; one shield with MAX6675 A/D converter, sufficient quantities of cables and connectors, equipment for making electrical circuits, one 100 mL beaker and a hotplate with a magnetic stirrer. This method was described by [9].

### 2.3 Calibration with a Hand-Held Calibrator (CL-477) by Omega Engineering

A 4–1/2 digit digital voltmeter; a precision mV/mA source (resolution: 1 μV/μA; accuracy: ±0.01% [±2 μV]); a recirculating ice bath (0.00 °C ± 0.05 °C); a type K thermocouple wire; a copper hookup wire; a potentiometer adjustment tool.

For temperature calibration, a recirculating ice bath, a type K thermocouple probe, and a precision millivolt/microvolt source are required. The probe is constructed as follows: With a suitable length of type K thermocouple wire and copper hookup wire, form junctions between each thermocouple wire (HI and LO) with a length of copper wire. Electrically isolate these junctions and insert both junctions in a recirculating ice bath (Fig. 2).

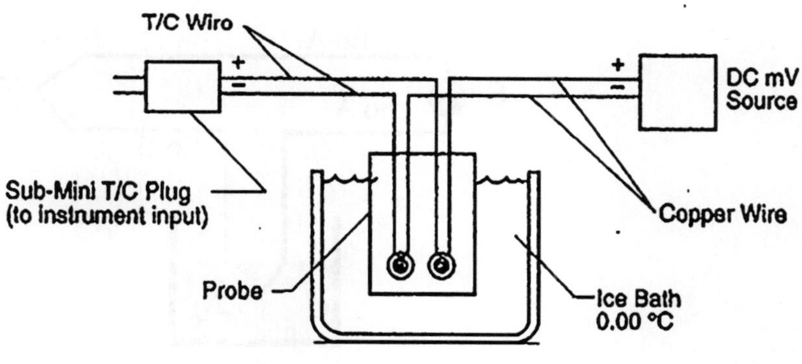


**Fig. 2** Calibration setup. (Source: OMEGA [10])

## 3 Methods

### 3.1 General Calibration Process

*STEP 1* Mount the system represented in Fig. 3. The junction of reference must remain at the same environmental temperature; the value will be determined before the process through a mercury thermometer.

*STEP 2* In the refractory container, heat 200 mL of water with the boiler until the mercury thermometer indicates a temperature between 90 °C and 100 °C.

*STEP 3* Dip the junction of the thermocouple measurement in the hot water. Measure the potential difference using the voltmeter and record the water temperature using the mercury thermometer.

*STEP 4* It is necessary to measure the potential difference in the thermocouple for various water temperatures. For that, add cold water, little by little, to the hot water contained in the refractory and repeat the measurement. The thermocouple and the thermometer bulb must be in the exact location, so place them close together.

*STEP 5* Create a graph of the potential difference in the thermocouple as a function of water temperature. Based on the graph, evaluate whether the Seebeck coefficient of this thermocouple remains constant within the observed temperature range. Therefore, make a linear regression of the measurement results and determine the value of this coefficient. Then, write the calibration equation $\varepsilon(T)$ of the thermocouple. Determine the value of reference temperature ($T_R$) by linear regression.

*STEP 6* After these steps, the thermocouple is calibrated and can be applied in different processes. To evaluate the operation of the device, measure the temperature of the environment and of people using a mercury thermometer, and compare the values obtained in both cases. The thermocouple and thermometer can be left closed.

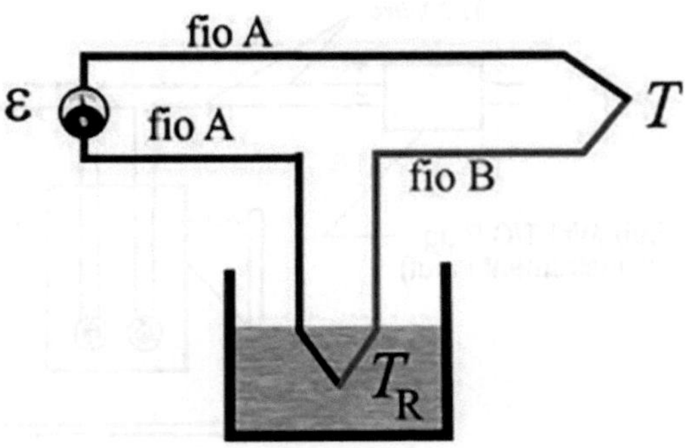


**Fig. 3** Schematic diagram of a thermocouple consisting of two wires (A and B) of different materials. (Source: Campos et al. [8])

### 3.2 Calibration of a Type K Thermocouple for Use in a Data Acquisition System with Arduino

| | |
|---|---|
| *STEP 1* | Mount the board A/D MAX6675A and connect the thermocouple. Connect the A/D MAX6675A in the Arduino and this to the computer. |
| *STEP 2* | Insert the thermocouple and the standard thermometer in a beaker with water. |
| *STEP 3* | Select the serial port. The maximum difference will be the maximum discrepancy between the thermometer and the thermocouple being calibrated. The calibration must be started along with heating and agitating the water. |
| *STEP 4* | The heating water is maintained at 9 to 80 °C, and the values must be compared using both the thermometer and the thermocouple. |
| *STEP 5* | Therefore, make a linear regression of the measurement results and determine the value of this coefficient. Then, write the calibration equation $\varepsilon(T)$ of the thermocouple, and correct the values of temperature read by the thermocouple in relation to the thermometer. |

### 3.3 Calibration with a Hand-Held Calibrator (CL-477) by Omega Engineering

#### 3.3.1 mV Calibration

| | |
|---|---|
| *STEP 1* | Remove the battery from the instrument (refer to Battery Replacement). Remove the bottom case half from the printed circuit board assemblies by extracting three screws from the case bottom and one screw holding the wrist strap lug. Remove the top case half from the printed circuit board assemblies by extracting the screw on the clad side of the Analog Board between the output controls. Reconnect the battery to the connector. |
| *STEP 2* | Measure voltage at TP-8 (+) on the display board (top board) with DVM common (−) at TP-2 on the analog board (bottom board). Voltage at TP-8 should be between 3.680 and 3.720 V. If the measured voltage is outside of the range, adjust R9 on the analog board for DVM reading at TP-8 of 3.695 V. |
| *STEP 3* | Set the instrument to indicator mode with an mV input. |
| *STEP 4* | Connect the precision mV/mA calibrator source to the T/C input, ensuring the proper polarity is observed. |
| *STEP 5* | Adjust the source output to 0.000 mV (short). |
| *STEP 6* | Select the high-resolution display (0.00 mV). The display should read 0.00 with a flashing minus sign. Turn R21 clockwise: display should read 0.00, 0.01, 0.02, 0.03. Turn R21 counterclockwise: display should read 0.03, 0.02, 0.01, 0.00, −0.00, −0.01, −0.02, −0.03. Adjust R21 for the display reading of 0.00 with a flashing minus sign. If readings are not stable, R26 on the analog board may need to be adjusted (normally adjusted near the center). |
| *STEP 7* | Adjust the source output to −7.000 mV. |

(continued)

| | |
|---|---|
| *STEP 8* | Adjust R28 on the analog board for an exact display reading of −7.000 mV. |
| *STEP 9* | Adjust the source output to 0.000 mV, check zero reading, and adjust R21 on the display board as required for a correct (0.00) reading. |
| *STEP 10* | Adjust the source output to +27.000 mV. |
| *STEP 11* | Adjust R9 on the analog board for an exact display reading of 27.00 mV. |
| *STEP 12* | Repeat steps 7 through 11 as necessary until no further adjustments are required for accurate display readings. |
| *STEP 13* | Switch instrument to low-resolution display (0.0) and verify the display readings of −7.0 mV, 0.0 mV, and 27.0 mV with an input of −7.00 mV, 0.000 mV, and 27.000 mV, respectively. |
| *STEP 14* | Perform temperature calibration as outlined in the following procedure. |

*3.3.2 Temperature Calibration*

| | |
|---|---|
| *STEP 1* | Set instrument to indicator mode, Type K range, 0.1° resolution, °F scale. |
| *STEP 2* | Observing proper polarity, connect the thermocouple wires from the ice bath to the instrument T/C input. |
| *STEP 3* | Observing proper polarity, connect the copper wires from the ice bath to the calibrator source. |
| *STEP 4* | Adjust the source output to 0.000 mV (short). Allow 15 min for stabilization. |
| *STEP 5* | Adjust R6 on the analog board for a display reading of 32.0 °F. |
| *STEP 6* | Adjust the source output to −4.513 mV. |
| *STEP 7* | Adjust R28 on the analog board for a display reading of −209.0 °F. |
| *STEP 8* | Recheck the display with zero input. Adjust R6 as necessary for a display reading of 32.0 °F. |
| *STEP 9* | Adjust source output to +7.848 mV. |
| *STEP 10* | Adjust R9 on the analog board for a display reading of 379.0 °F. |
| *STEP 11* | Repeat steps 6 through 10 until no further adjustments are required for correct display readings. |
| *STEP 12* | Set the instruments to indicator mode with an mV input. |
| *STEP 13* | Check the display reading with the following inputs: −7.000 mV (−7.000 mV display), 0.000 mV (0.00 mV display), + 27.00 mV display. All readings should be within a tolerance of ±0.01 mV. If not, the entire calibration procedure must be repeated. |
| *STEP 14* | Perform mA input calibration as outlined in the following procedure. |

*3.3.3 mA Input Calibration*

| | |
|---|---|
| *STEP 1* | Observe proper polarity and connect the precision mA source to the mA input jacks of the instrument. |
| *STEP 2* | Set the instrument to mA input mode. |
| *STEP 3* | Set the mA A/B switch to the "A" position. |
| *STEP 4* | Adjust the source output to 0.000 mA. |
| *STEP 5* | Adjust R5 on the display board for a display reading of 0.00 mA. |
| *STEP 6* | Adjust the source output to 25.000 mA. |
| *STEP 7* | Adjust R20 on the analog board for a display reading of 25.00 mA. |
| *STEP 8* | Repeat steps 4 through 7 until no further adjustments are required for correct display readings. |
| *STEP 9* | Carefully reassemble the instrument. |
| *STEP 10* | Calibration completed. |

## 4 Notes

Calibration of equipment is crucial in any process. For the thermocouples, it would be no different. It is realized that calibration is necessary for obtaining a device with a degree of data certainty. Currently, many methods are being developed for various applications. These characteristics are not different for the thermocouple. Therefore, the selection of an ideal calibration method is related to the available mechanisms in the industry and laboratory. Therefore, the methods presented in this chapter represent many other methods available in different sources. The primary objective is for the reader to gain a general understanding of the basic and intermediate methods and to present the easy method of calibration so that it is possible to select the most suitable method for their needs.

## References

1. Seebeck TJ (1822) Abhandlungen der Deutschen Akademie der Wissenschaften. Berlin 265
2. Mc Cullagh JJP, Setchell DJ, Gulabivala K, Hussey DL, Biagioni P, Lamey P-J et al (2000) A comparison of thermocouple and infrared thermographic analysis of temperature rise on the root surface during the continuous wave of condensation technique. Int Endod J 33:326–332. https://doi.org/10.1046/j.1365-2591.2000.00302.x
3. Ripple DC, Garrity KM (2006) Uncertainty budgets for comparison calibrations of thermocouples. J Measure Sci 1:28–34
4. Wu J (2023) A basic guide to thermocouple measurements. Texas Instruments Incorporated, pp 1–36
5. ANSI (2018) Thermocouples: Calibration tables, types, specifications, and standards. https://blog.ansi.org/2018/10/thermocouples-calibration-table-ansi-mc961/

6. EURAMET (2020) Guidelines on the calibration of thermocouples
7. TEMPCO (2022) Thermocouple calibration type selection guide. https://www.tempco.com/Tempco/Blog/Thermocouple-Calibration-Type-Selection-Guide.htm#:~:text=According%20to%20the%20American%20National,types%20being%20J%20and%20K
8. Campos AA, Alves ES, Speziali NL (2018) Física experimental básica na universidade. Ed. UFMG
9. Romio C, Lorscheiter TA, Silva INL (2012) Calibration of a type k thermocouple for use in a data acquisition system with the arduino. Periódico Tchê Química 09:60–66. https://doi.org/10.52571/PTQ.v9.n17.2011.60_Periodico17_pgs_60_66.pdf
10. OMEGA (2024) Omega Engineering Inc. https://assets.omega.com/manuals/M0196.pdf

# Chapter 12

# Temperature Measurements Using Thermocouples

**Matheus S. Mourão and Roberto Pessanha da Silva Pires**

**Abstract**

Temperature measurement is an essential tool in process control. The industrial food process has temperature as an important process variable. In this sense, this chapter presents many forms of temperature measurements using thermocouples. Technicians widely use thermocouples in various measurement processes for their ease of reading and simple composition. Beyond applications in specific processes, this chapter presents a general process for developing thermocouple applications for multiple purposes. Altogether, the application of thermocouples, combined with software, is widespread because it simplifies the process, yields more accurate results, and facilitates better data analysis. Furthermore, the automatization process is a demand of modern times, and the application of models in different software simplifies the data acquisition process.

**Key words** Cold-junction compensation, Food processing, Industrial technology, Evaluation technology

## 1 Introduction

Temperature measurements are essential and critical in different processes, like food processing, industrial material production, and other industrial applications related to industrial technology. Ensuring process efficiency and product consistency is crucial in industrial processing, as well as maintaining precise temperature control [1]. Therefore, safe temperature measurement on the International Temperature Scale with high reproducibility and low measurement uncertainty is necessary for ideal processing [2]. In the food production chain, temperature plays a crucial role in controlling microbial growth. However, microbial pathogens, such as viruses, parasites, and bacteria, remain the primary contaminants associated with diseases caused by food poisoning [3].

The contamination of food matrices can occur during the production, transportation, and consumption stages of food. Among the most common foodborne pathogens, *Listeria*, *Escherichia*, *Campylobacter*, and *Salmonella* are encountered [3]. Food

Anderson S. Sant'Ana (ed.), *Heat Resistant Microorganisms in Food Safety and Spoilage*, Methods and Protocols in Food Science, https://doi.org/10.1007/978-1-0716-5280-0_12,

scientists use thermal treatment techniques to decontaminate food matrices, where temperature acts as both a control parameter and an evaluation technology to assess microbial inactivation. Heat inactivation can affect both vegetative and dormant cells, favoring the loss of activity [4]. However, thermal treatment can affect the material properties of food, such as reducing the hardness of solid foods and the viscosity of fluid foods. The temperature can also favor the release of flavor components [5, 6]. Therefore, the application of devices for evaluating temperature is important in food processing for obtaining a final product with the desired features.

Thermocouples are widely used temperature sensors in different applications. Different types of thermocouples accommodate a range of mechanisms for temperature application and measurement. Different studies have applied this device in sensory perception [7], the evaluation of temperature in the combustion chamber and flue gas duct [8], and monitoring of the alcoholic fermentation process [9]. In the industrial process of food formulation, the techniques of thermal inactivation are applied to decontaminate different food matrices. These devices exhibit a dynamic effectiveness in the process, which can be evaluated in terms of the Z and D values [4].

Nevertheless, periodic outbreaks continue to be observed in various food matrices worldwide. These problems can be related to the characteristics of the thermocouples used. Structural changes in the wires of the devices can occur due to thermal stress and external mechanical influences that disfavor accurate temperature measurement, facilitating this error. Knowing the process is crucial for selecting the ideal thermocouples that possess the optimal characteristics for the application [2, 4].

## 2 Material

### 2.1 Evaluation of the Reduction of Salmonella Surrogate in Beef Strip Loins at Temperatures Lower than 54.4 °C

Rub the ingredients and raw meat materials together, including a cocktail of *Escherichia coli*, a trypticase soy broth (TSB), a brine solution, a cook-in bag, and a type K thermocouple. This method was described by [10].

### 2.2 Monitoring the Alcoholic Fermentation Process Based on Acoustic Emission Analysis

Mix 8 L of a solution of brown sugar cane and distilled water, and the pH of the solution should be fixed at 3.8 by applying citric acid; an air flow; an experimental reactor of 20 cm in diameter and 40 cm in height made of acrylic to perform regular visual inspections during the fermentation process; a Teledyne RESON ® TC-4033 hydrophone with the frequency range: 1 Hz –144 kHz and sensitivity: 1 V/μPa; a type J thermocouple; a G.R.A.S.® Power

Module; MATLAB® software; NI USB 6211 (0–100 kHz) data acquisition card. The RMS of the signal will be calculated applying the Parseval theorem (Eq. 1).

$$\sum_{n=1}^{N} x^2[n] = \frac{1}{N}\sum_{m=1}^{N} |X[m]|^2 \quad (1)$$

where $X[m]$ is the discrete time Fast Fourier Transform (FFT) of $x[n]$ and $N$ is the sample size. This method was described by [9].

### 2.3 Application of a Thermocouple with Software Compensation

Thermocouple (different types), National Instruments NI-DAQ driver software, as well as National Instruments LabVIEW or Measurement Studio, and direct-reading sensor measures. This method was described by [11].

## 3 Methods

### 3.1 Evaluation of the Reduction of Salmonella *Surrogate in Beef Strip Loins at Temperatures Lower than 54.4 °C*

| | |
|---|---|
| *STEP 1* | Inoculations should be prepared by inoculating TSB with a strain of *E. coli* and allowing it to grow at 37 °C for approximately 24 h. |
| *STEP 2* | Strip loins should be dip inoculated with *E. coli* to achieve a 7.5-$\log_{10}$ CFU/g inoculation level on the meat. |
| *STEP 3* | Strip loins should be pumped 15% with a brine solution and then rubbed with the rub. |
| *STEP 4* | They should then be placed into cook-in bags and vacuum-sealed. Packages will be placed on a smokehouse trolley in the smokehouse. |
| *STEP 5* | The combination of temperatures and times held should be 54.4 °C for 2 and 3 h, 51.7 °C for 3 and 5 h, and 48.9 °C for 10 and 12 h. times should be determined utilizing a model from the North American Meat Institute. |
| *STEP 6* | Internal temperatures should be continuously monitored utilizing type K thermocouples. Once removed from the smokehouse, 1 kg samples should be taken from each strip loin and vacuum-packaged for *salmonella* surrogate enumeration. |
| *STEP 7* | Samples should be taken to food safety net services for enumeration. MacConkey Sorbital agar will be used to determine the survival of salmonella surrogates. |
| *STEP 8* | The experiment should consist of three replications, with two samples per treatment per replication. |

### 3.2 Monitoring the Alcoholic Fermentation Process Based on Acoustic Emission Analysis

| | |
|---|---|
| *STEP 1* | Two liters of the solution should be separated from the prepared solution, and 3 g of dry yeast (*Saccharomyces cerevisiae*) should be inoculated. Ammonium phosphate should be used as a nitrogen substrate. |
| *STEP 2* | An airflow of approximately 50 L/min (1 atm/25 °C) should be vigorously injected into 2 L of the inoculated sugar solution. The air should be uniformly distributed in a bubble curtain from the bottom by applying a specifically designed air diffuser system with an average bubble size of 0.8 mm. |
| *STEP 3* | The aerobic stage of the inoculated solution should be maintained for 6 h to ensure convenient microbial biomass growth, reaching a yeast concentration of 6 g/L. |
| *STEP 4* | After the aerobic stage, the 2 L enriched in yeast cells (biomass) should be inoculated (mixed) into the other 6 L of brown sugar solution to complete the operational capacity (8 L) of an experimental reactor of 20 cm in diameter and 40 cm in height made of acrylic to perform regular visual inspections during the fermentation process. |
| *STEP 5* | To acquire the acoustic signal produced by the $CO_2$ bubbles during the fermentation process, a Teledyne RESON TC-4033 hydrophone should be used, with a frequency range of 1 Hz to 144 kHz and a sensitivity of 1 V/μPa. The hydrophone should be introduced into the fermenting liquid at a depth of about 7 cm, as depicted. |
| *STEP 6* | The reactor should be covered with thermal insulation foam to minimize heat loss to the environment. Thermal isolation enables the intentional increase in temperature of the fermenting liquid in the reactor, according to the fermentation kinetics. |
| *STEP 7* | The thermocouple should be placed at a close distance (approximately 2 cm) from the hydrophone sensor to take acoustic and temperature measurements at nearly the same point. Both the hydrophone and the thermocouple should be placed in the central region of the reactor to avoid possible interferences in sound and temperature measurements due to the influence of geometry and other effects, such as bubbles colliding with the reactor wall or lower temperature gradients formed nearby. |
| *STEP 8* | The signals from the hydrophone and the thermocouple are digitized using a NI USB 6211 (0–100 kHz) data acquisition card to be recorded by the computer using dedicated software programmed on LabView®. The acoustic data should be processed using MATLAB® software. |

(continued)

| | |
|---|---|
| *STEP 9* | To eliminate any external interference associated with the frequency of interest, a spectrogram and the components of frequency within the range of 0.2–100 kHz were recorded in the reactor filled with water. No external interferences (noise) should be found in the original signal in the selected frequency range. A high-pass (HP) filter should be applied for the signal filtering in the range of interest (0.2–10 kHz). |
| *STEP 10* | The NI USB-6211 acquisition card is designed for fast settling times at high scanning rates, ensuring 16-bit accuracy (absolute accuracy at full scale = 2.69 μV). The G.R.A.S. Power module type 12AQ has a gain error of ±0.1 dB. To minimize errors, the measurement chain should be calibrated using a G.R.A.S® 42AP intelligent pistonphone. |
| *STEP 11* | Different descriptors of the fermentation acoustic signal acquired by the hydrophone should be evaluated as a function of time, such as amplitude, frequency distribution, and the root-mean-square (RMS) power of the signal. Additionally, the temperature of the fermenting liquid and the room temperature should be recorded simultaneously in the time domain. |
| *STEP 12* | The RMS of the signal should be calculated by applying Parseval's theorem (with the contribution of Plancherel and Rayleigh), which considers the energy conservation law and is recommended for nonperiodic signal analysis, which is often described for discrete time signals with Eq. 1. |
| *STEP 13* | After 4 days, the fermentation will be finished, and the temperature and acoustic parameters will exhibit an asymptotic behavior in the time domain, thus returning to the initial values recorded at "time zero". Fermentation experiments should be replicated eleven times for further mathematical processing and to ensure the reproducibility of the acquired data. |

## 3.3 Application of a Thermocouple with Software Compensation

### 3.3.1 Procedure 1—Direct Voltage Addition Method for Software Cold-Junction Compensation

| | |
|---|---|
| *STEP 1* | Initially, apply Eq. (2) to find the accurate open-circuit voltage that the thermocouple would produce with a reference junction at 0 °C: $V_{TC}(V_{TC}) = V_{MEAS} + V_{TC}(V_{ref})$ (2) |
| *STEP 2* | Measure the reference-junction temperature, $T_{ref}$. |

(continued)

| | |
|---|---|
| *STEP 3* | Convert this temperature into an equivalent voltage for the thermocouple type undergoing testing, $V_{TC}(T_{ref})$. Either the NIST reference tables or polynomials that assume a reference junction at 0 °C can be used. |
| *STEP 4* | Add this equivalent voltage to the measured voltage, $V_{MEAS}$, to obtain the accurate open-circuit voltage that the thermocouple would produce with a reference junction at 0 °C, $V_{TC}(T_{TC})$. |
| *STEP 5* | Convert the resulting voltage into a temperature; this value is the thermocouple temperature, $T_{TC}$. |

This compensation method requires translating the junction temperature into a thermocouple voltage, followed by translating the new voltage into a corresponding temperature. Each of these translation steps requires either a polynomial calculation or a look-up table.

#### *3.3.2 Procedure 2—Temperature Addition Method for Software Cold-Junction Compensation*

A second, more straightforward software compensation approach leverages the fact that thermocouple output voltages are approximately linear over small temperature deviations. Therefore, for minor deviations in temperature, Eq. (3) can be used:

$$V_{TC}(T_1) - V_{TC}(T_2) \approx V_{TC}(T_1 - T_2) \tag{3}$$

This assumption is valid if $T_1$ is relatively close to $T_2$ because the thermocouple voltage-versus-temperature curve is approximately linear for minor temperature variations. Assuming that the thermocouple temperature is relatively close to the reference temperature, Eq. 3 can be rewritten as Eq. 4:

$$V_{MEAS} = V_{TC}(T_{TC} - T_{ref}) \tag{4}$$

If standard NIST thermocouple reference tables or equations are used, the voltage $V_{TC}$ is a function of temperature relative to a reference temperature of 0 °C. By assuming linearity and using Eq. 4, it can be assumed that the voltage-versus-temperature curves with a reference temperature of 0 °C are identical to those with a reference temperature of $T_{ref}$. Therefore, we can convert the measured voltage into a temperature using the NIST reference tables. This temperature is the difference between the temperatures TTC and *T*ref.

| | |
|---|---|
| *STEP 1* | Measure the reference-junction temperature, $T_{ref}$. |
| *STEP 2* | Convert the measured voltage, $V_{MEAS}$, into a temperature using the voltage-to-temperature relationship of the thermocouple. This temperature is approximately the difference between the thermocouple and the cold-junction reference, $T_{TC}$—$T_{ref}$. |
| *STEP 3* | Add the temperature of the reference junction, $T_{ref}$, to this value. This is the thermocouple temperature. |

## 4 Conclusions

There are various forms of temperature measurement. In this chapter, various forms of thermocouple applications in food processing are presented. In these processes, the thermocouple is used as a mechanism for determining temperature throughout the entire process or a specific part of it. It is essential to be aware of temperature variations so that necessary corrections can be made. In cases where the action of the thermocouples is the principal factor, it is possible to apply methods that result in data of interest. The thermocouple is a versatile device that can be applied to various methods of temperature measurement. That way, it is not difficult to find various research works that apply this device in different types of studies.

### References


1. Jagtap S, Rane S, Gosavi S (2016) Synthesis, characterization and fabrication of NTC thick film thermistor using lead free glass frit. J Mater Sci Eng A 6. https://doi.org/10.17265/2161-6213/2016.11-12.003
2. Edler F (2023) Reliable and traceable temperature measurements using thermocouples. Johnson Matthey Technol Rev 67:65–76. https://doi.org/10.1595/205651323X16692809325480
3. WHO (2024) Food safety. https://www.who.int/news-room/fact-sheets/detail/food-safety
4. Harguindeguy M, Gómez-Camacho CE (2021) Pulsed light (PL) treatments on almond kernels: salmonella enteritidis inactivation kinetics and infrared thermography insights. Food Bioprocess Technol 14:2323–2335. https://doi.org/10.1007/s11947-021-02725-9
5. Bozdogan A (2015) Viscosity behavior of bitter orange (*Citrus aurantium*) juice as affected by temperature and concentration. CyTA J Food:1–6. https://doi.org/10.1080/19476337.2015.1012120
6. Schoumacker R, Martin C, Thomas-Danguin-T, Guichard E, Le Quéré JL, Labouré H (2017) Fat perception in cottage cheese: the contribution of aroma and tasting temperature. Food Qual Prefer 56:241–246. https://doi.org/10.1016/j.foodqual.2016.04.003
7. Lv C, Wang X, Chen J, Yang N, Fisk I (2019) A non-invasive measurement of tongue surface temperature. Food Res Int 116:499–507. https://doi.org/10.1016/j.foodres.2018.08.066
8. Najser T, Gaze B, Knutel B, Verner A, Najser J, Mikeska M et al (2022) Analysis of the effect of catalytic additives in the agricultural waste combustion process. Materials 15:3526. https://doi.org/10.3390/ma15103526
9. Mamolar-Domenech S, Crespo-Sariol H, Sáenz-Díez JC, Sánchez-Roca A, Latorre-Biel J-I, Blanco J (2023) A new approach for monitoring the alcoholic fermentation process based on acoustic emission analysis: a preliminary assessment. J Food Eng 353:111537. https://doi.org/10.1016/j.jfoodeng.2023.111537
10. Mendes B, Krage E, Henson J, Mckeith AG (2017) Evaluation of the reduction of salmonella surrogate in beef strip loins at temperatures lower than 54.4°C. Meat Muscle Biol 1:132–132. https://doi.org/10.22175/rmc2017.125
11. Ferguson T (2011) Measuring temperature with thermocouples—a tutorial

# Chapter 13

# Performing Heat Penetration Studies in Commercially Sterile Foods

Larissa Pereira Margalho and Eugénio da Piedade Edmundo Sitoe

**Abstract**

Heat penetration studies are essential in developing and validating thermal processes for food safety and quality assurance in commercial sterilization and pasteurization. These studies ensure that the thermal treatment delivers the minimum lethality required to destroy pathogenic and spoilage microorganisms, particularly *Clostridium botulinum*. This study presents a step-by-step protocol for conducting heat penetration studies, focusing on accurate data collection and analysis to determine thermal processing efficacy. The protocol includes selecting appropriate test food matrices, preparing temperature measurement probes, defining target cold spots within the product, and configuring data acquisition systems. The heating and cooling cycles are carefully monitored under controlled conditions to evaluate temperature distribution and heat transfer dynamics. Key parameters such as F-values, lethality rates, and come-up time are obtained based on time–temperature profiles. This methodology ensures reproducibility, minimizes variability, and provides insights for optimizing thermal processing parameters. The findings improve process validation, ensuring regulatory compliance and microbial safety while maintaining product quality. This protocol is critical for food scientists, process engineers, and regulatory professionals in designing and validating thermal food processing operations.

**Key words** Heat penetration, Commercial sterilization, Thermal lethality, Process validation, Food safety

## 1 Introduction

Thermal processing, including sterilization and pasteurization, is a cornerstone of the food industry, ensuring microbial safety and extending the shelf life of heat-processed foods. The efficacy of these processes hinges on the uniform delivery of heat to the slowest-heating region of the product, known as the cold spot. Heat penetration studies are indispensable for understanding heat transfer dynamics, validating process efficacy, and ensuring compliance with stringent regulatory standards for food safety [1, 2]. These studies provide critical time–temperature data, which are used to calculate key parameters such as F-values, lethality

Anderson S. Sant'Ana (ed.), *Heat Resistant Microorganisms in Food Safety and Spoilage*, Methods and Protocols in Food Science, https://doi.org/10.1007/978-1-0716-5280-0_13,

rates, and come-up time, all essential for achieving the desired microbial inactivation while preserving product quality [3].

The growing complexity of modern food products, including heterogeneous matrices and ready-to-eat meals, has underscored the need for precise and reproducible heat penetration studies. The location of the cold spot can vary significantly depending on factors such as product composition, container geometry, and processing conditions [4]. Accurate identification and monitoring of the cold spot are therefore critical to ensuring uniform heat distribution and preventing under-processing, which can compromise food safety, or over-processing, which can degrade sensory and nutritional qualities.

Despite their importance, heat penetration studies present several challenges, including variability in heat transfer rates, the need for precise temperature measurement, and the complexity of data analysis. Recent advancements in data acquisition systems, such as wireless thermocouples and real-time monitoring tools, have significantly improved the accuracy and efficiency of these studies [5]. Additionally, computational modeling and machine learning approaches are emerging as powerful tools for predicting heat transfer dynamics and optimizing thermal processes [6].

This chapter presents a step-by-step protocol for performing heat penetration studies in heat-processed foods. The protocol includes the selection of representative food matrices, calibrating and strategically placing temperature measurement probes, identifying target cold spots, and configuring advanced data acquisition systems [7, 8]. By providing actionable insights for optimizing thermal processing parameters, this protocol advances food safety and quality assurance in the food industry. Furthermore, it aligns with global regulatory requirements and supports the development of innovative thermal processing technologies to meet the growing demand for safe, high-quality food products.

## 2 Materials

1. Product samples (formulated per commercial standard).
2. Packaging containers (cans, glass jars, pouches, semi-rigid trays, cartons, etc.).
3. Thermocouples (Needle, Type T or K; diameter 0.2–0.5 mm).
4. Data acquisition system (logging at 1–2 s intervals).
5. Retort (static, rotary, water spray, or steam).
6. Calibration device for thermocouples (using an ice bath and boiling water).
7. Temperature monitoring system with software for data logging.

8. Sealing equipment (for sealing cans or containers).
9. Safety equipment (gloves, goggles, and lab coat).
10. Heat-resistant sealants (e.g., silicone and epoxy).
11. Drilling tools, guides, and positioning aids.

## 3 Methods

### 3.1 Instrumented Container and Thermocouple Preparation

1. Before starting a heat penetration study, it is necessary to characterize the container type (metal cans, glass jars, retort pouches, and semi-rigid containers), recording its size and dimensions.
   (a) For rigid containers, quantitative measurements of internal vacuum and headspace volume must be recorded.
   (b) For flexible and semi-rigid containers, the residual gas volume should be determined through volumetric or manometric analysis. Entrapped gases may form an insulating boundary layer within the container, leading to displacement of the thermal cold spot and reduction in heat transfer efficiency. Empirical studies indicate that applying controlled overpressure during thermal processing mitigates these effects by suppressing gas expansion and improving thermal conduction.
      b.1. Pouches should be analyzed with respect to their physical properties: total thickness, overall migration residue, internal burst strength for seal integrity, heat seal strength, process resistance test, tensile strength and elongation at break, and residual air in the pouch.
2. Then, it is necessary to make instrumented containers, by making a small hole with a needle, a puncher, or an electric drill. It will depend on the type of containers, thermocouples, and food to be analyzed. It is important to do this away from the seams or seals of the packaging container.
3. The entry points of thermocouples must be hermetically sealed to prevent gaseous or fluid leakage during thermal processing. Effective sealing can be accomplished through the use of compression fittings or packing glands, incorporating elastomeric gaskets (e.g., rubber or silicone-based polymers), with fastening nuts securely torqued to ensure a pressure-resistant seal on the containment vessel.
4. Prepare thermocouples to measure the internal temperature of the food. Typically, thermocouples are inserted into the geometric center of the product, where heating is slowest. In this case, it is necessary to consider the ratio of liquid/solids, the headspace, the geometry of the container to determine the cold

spot, and the type of heat penetration rates (if by conduction, convection, or both).

5. If needed, insert additional thermocouples at various points (e.g., near the container's walls or lid) to monitor temperature gradients, mostly in convection heating transfer.

### 3.2 Setting the Container's Cold Spot

Thermocouple positioning within the product matrix must be optimized to ensure accurate thermal monitoring, with particular emphasis on the identification and validation of the thermal cold spot or zone of slowest heating in a container. In this case, it is assumed that all other points in the container receive more heat and are therefore adequately processed. The cold spot is experimentally determined using multiple thermocouples in trial containers. Improper placement leads to under- or overestimation of lethality, compromising food safety or quality. Take **Notes 1–3** in consideration.

The cold spot is typically at the geometric center of the can for conduction-heated foods (e.g., mashed potatoes, high-viscosity foods) when the container has a height: diameter ratio between 0.95 and 0.3 [9] and approximately 10–15% above the base for low-viscosity or convective fluids (e.g., soup, chunky vegetables) during still processing. Consider **Note 4** for any change in containers.

#### 3.2.1 Thermocouple Positioning by Container Type

**Metal Cans**

1. Mark the external can wall at the location corresponding to the cold spot.
2. Preferably insert through the side wall or bottom, avoiding the top lid if sealed by a double seam.
3. Carefully drill a hole with a bit slightly smaller than the thermocouple diameter (1.5–2 mm).
4. Deburr the edges to avoid damaging the wire or compromising sealing.
5. Clean the area to remove metal shavings (important for aseptic conditions).
6. Insert the thermocouple so that the sensing tip is positioned exactly at the cold spot.
7. Use a depth gauge or calibrate using a transparent replica of the product (e.g., gel model).
8. Use a compression fitting or bulkhead feedthrough to ensure a hermetic seal.
9. Apply a high-temperature silicone sealant around the insertion point.
10. Let it cure fully before processing (typically 12–24 h at room temperature).

11. For extra security in high-pressure retorts, use epoxy over the fitting.
12. Gently press on the can to ensure no leakage.
13. Optionally, immerse the can in water and look for bubbles (leak test).

**Glass Jars**

1. Fill the jar with the target food product under standard operating conditions.
2. Leave adequate headspace per process standard (typically 5–10%).
3. Determine the cold spot, based on product characteristics and prior studies. For conduction: exact center of the food mass; for convection: ~ 1/tenth jar height from the base.
4. Use a drill or punch to make a 1.5–2 mm hole in the center or offset side of the metal lid.
5. Never drill into the glass body due to high fracture risk.
6. Deburr and clean the whole area thoroughly.
7. Feed the thermocouple through the drilled hole. Position the sensing tip precisely at the cold spot using a calibrated depth gauge, ruler, or transparent gel model. Secure the position with a stopper or fitting to avoid shifting during processing.
8. Use a compression fitting, bulkhead connector, or high-temperature silicone sealant around the thermocouple where it exits the lid. Ensure that the seal is Hermetic (prevents leakage during processing), thermally stable (resistant to $\geq$130 °C), food-safe, and inert.
9. Gently screw the lid onto the jar without overtightening, allowing for pressure equalization during retort heating.
10. Ensure the thermocouple is not pulling upward or compressing the lid seal unevenly.
11. Submerge the jar in water and apply vacuum to check for leaks (no bubble formation).

**Flexible Pouches**

1. Place the pouch on a flat surface in the intended processing orientation.
2. Mark the external point where the thermocouple should reach the internal cold spot.
3. Clean and dry the surface.
4. Use a heated stainless-steel needle or a pointed puncture probe (~0.8–1 mm diameter) to gently melt and pierce through the

film. Use a temperature-controlled soldering iron tip if available. This creates a clean, resealable hole by melting rather than tearing the film layers.

5. Apply food-grade silicone generously at the wire/film interface on the outside.
6. Cover the site with a retortable adhesive patch, pressing firmly for full contact.
7. Allow full curing of the sealant: 12–24 h at room temperature (as per silicone/adhesive specification) before thermal processing.
8. Perform a leak testing: Submerge the sealed pouch in water, gently applying pressure. Inspect for bubbles at the insertion site, indicating leakage.

**Semi-rigid Trays**

1. Determine and mark the cold spot based on product type and tray geometry. Mark the corresponding location on the lid or side wall of the tray.
2. Select the insertion point. The preferred site is usually the top lidding film (center or offset), as it allows minimal damage, can be easily resealed, and avoids interference with the sealing rims or tray corners.
   (a) Alternate sites include tray sidewalls, but sealing is more complex due to the curvature and material thickness.
3. Puncture the lidding film using a sterile, heated puncture probe (e.g., 1 mm stainless-steel needle) or a micro drill.
4. Gently melt through the lidding film with a hot probe to create a clean puncture, avoiding delamination or tearing.
   (a) In the case of aluminum foil lids, use a micro drill or a fine puncture awl.
5. Insert the thermocouple immediately through the puncture. Advance carefully to the cold spot using a depth gauge or positioning template. Use a probe guide if needed to stabilize the wire.
6. Seal the thermocouple entry site applying food-grade high-temperature silicone at the thermocouple–lid interface.
7. Cover the insertion point with a retortable adhesive patch, pressing firmly to ensure adhesion.
   (a) Optionally, use a foil-backed adhesive sealing disk or bulkhead grommet for high-pressure applications.
8. Allow the sealant to cure fully (12–24 h, as per product instructions).

9. Perform a leak testing (Optional) by submerging the tray in clean water. Apply gentle pressure to detect air bubbles from the insertion site. If leakage is observed, re-seal and retest.

### 3.3 Sample Preparation

1. Select the food product to be studied and package it in the instrumented container (e.g., cans, jars, pouches, etc.).
2. Fill containers consistently to ensure that product mass, density, and headspace are similar to what would be used in commercial production. It is important to weigh the amount of solids and liquids as well.
3. The solid content of heterogeneous products should be quantified both pre- and post-processing. The particulate fraction retained on a sieve must be gravimetrically analyzed and reported as a mass percentage relative to the total sample weight. In the case of solid food, measure the size and shape of the largest particulates/pieces. It will be important to use a small particle to impale the thermocouple probe at the cold spot.
4. The rheological properties (consistency and viscosity) of semi-liquid or liquid formulations should be quantitatively assessed both pre- and post-processing.
5. Ensure that product preparation prior to filling simulates commercial practice.
6. Filling temperature of the product should be controlled. It will affect the initial temperature which may influence some heat-penetration parameters.
7. Place the thermocouple in the cold spot of the product. It could require extra utensils, mostly in the case of pouches, as described in Subheading 3.2.1.
8. Seal containers properly to avoid leaks or contamination. Closing or sealing equipment should ensure a robust, hermetic seal that remains intact throughout the thermal processing operation.
9. Make a leak test submerging the container in water to check for bubbles escaping.
10. Test/instrumented containers should be placed at the slowest heating location in the retort, as determined by temperature and heat transfer distribution studies.

### 3.4 Retort Loading and Instrumentation

1. Choose the appropriate thermal processing equipment (retort, oven, water bath) based on the type of food and process (pasteurization, sterilization, etc.).

2. Preheat the processing equipment to the desired temperature (as specified by the target process).
3. Ensure that the thermocouples or data loggers are properly connected and ready to record the temperature in real time.
4. Load containers to replicate commercial stack density. Number the samples and the position in the equipment.
5. Distribute retort thermocouples to monitor the chamber temperature.
6. Connect all sensors to a calibrated data logger.
7. Start the thermal processing cycle according to the desired temperature and time. This could be batch or continuous processing.
8. Monitor and record the temperature data throughout the process. The key variables to monitor include:
   - $T_L$ = Heating medium temperature (e.g., steam, water, or air).
   - $T_g$ = Product center temperature.
   - $t_{cut}$ = Time to reach the target temperature.
   - $t_p$ = Time of process when the target temperature is achieved.
   - Heating curves = Time–temperature profile during the hold phase.
   - Cooling curves = Time–temperature profile during the cooling phase.

Consider **Notes 5–10** to ensure data reproducibility and quality, as well as analyst safety.

## 4 Notes

1. Before starting the heat penetration test, it is necessary to conduct a heat distribution test inside the autoclave, in order to guarantee a uniform distribution of heat and get the cold spot inside the equipment. At this point, the heat penetration tests will be conducted. Refer Chap. 17.
2. A DT retort study should be performed every 3 years or in the following cases: Relocation of the retort or installation of another retort that uses the same services; modification to the steam, water, or air supply; replacement of the key components, e.g., pumps and valves; repair or modification to water circulation or steam injection systems within the retort; and if there are any doubts about the performance of the retort.
3. Before starting the tests, calibrate the thermocouples (refer Chap. 11).

4. The type, geometry, and dimensions of the containers should replicate those employed in the intended process to ensure experimental validity and reproducibility. It also applies to the composition of foods. Any changes in those components requires a new validation study [8].
5. It is necessary to have all data related to the target microorganism for the evaluated food, as the z-value, and $D_0$ [7].
6. Tests should be repeated as many times as necessary for greater reproducibility [8].
7. Quantitative data on variability should be collected where applicable, and all relevant information should be thoroughly documented to facilitate a comprehensive understanding of potential variations in the heat penetration behavior [8].
8. In case of any change throughout the process, such as formulation, design and operation of autoclaves, changing of pumps, etc., a new heat penetration study must be carried out [8].
9. All instrumentation employed must exhibit high precision and appropriate scalability while being available in sufficient quantities to facilitate robust and reproducible analysis, thereby ensuring accurate and safe monitoring of the process environment [8].
10. Always wear appropriate protective gears (gloves and goggles) when working with hot equipment or during the handling of thermally processed food [7].

## Acknowledgments

L. P. Margalho and E.P.E. Sitoe acknowledge the financial support of the "Fundação de Amparo à Pesquisa do Estado de São Paulo" (FAPESP): Grant #2020/14631-6 and #2024/20257-0. This study was financed in part by the "Coordenação de Aperfeiçoamento de Pessoal de Nível Superior" (CAPES)—Finance Code 001.

### References

1. Downing DL (2013) A complete course in canning and related processes: processing procedures for canned food products. Elsevier
2. Sahoo M, Titikshya S, Aradwad P, Kumar V, Naik SN (2022) Blanching, pasteurization and sterilization: principles and applications. In: Thermal food engineering operations, pp 75–115
3. Holdsworth SD, Simpson R, Barbosa-Cánovas GV (2008) Thermal processing of packaged foods (Vol. 284). New York: Springer
4. Tucker G, Featherstone S (2021) Essentials of thermal processing. Wiley
5. Raits E, Pinte L, Kirse-Ozolina A, Muizniece-Brasava S (2021) A case-study: temperature

distribution and heat penetration in steam-air retort, using glass jars and retort pouches. 46:90
6. Godasiaei SH, Kamali HA (2025) Evaluating machine learning as an alternative to CFD for heat transfer modeling. Microgravity Sci Technol 37(1):1–14
7. IFTPS – Institute For Thermal Processing Specialists (2014) Guidelines for conducting thermal processing studies. Iftps, 1–45. http://iftps.org/wp-content/uploads/2017/12/Retort-Processing-Guidelines-02-13-14.pdf
8. Pflug IJ (2010) Procedures for carrying out a heat penetration test and analysis of the resulting data. Department of Food Science and Nutrition, University of Minnesota
9. Fellows PJ (2022) Food processing technology: principles and practice. Woodhead publishing

# Chapter 14

# Protocol for Estimating Heat Penetration Parameters of Foods

**Larissa Pereira Margalho and Eugénio da Piedade Edmundo Sitoe**


## Abstract

This protocol aims to develop a comprehensive method for estimating heat penetration parameters in food processing, focusing on the heating rate index ($f$) and lag factor ($j$), as well as the $F0$ calculation, which represents the sterilization value or lethality of a thermal process. In commercial sterilization, $F_0$ values are used to determine the necessary processing time to achieve the desired level of microbial inactivation. These parameters are crucial for determining the thermal behavior of foods during thermal processing, ensuring safety and quality. The protocol will incorporate various methods, including thermocouple placement and computational modeling, to accurately measure and predict heat penetration in different food types and packaging. It will address challenges such as variability in thermal properties and the influence of container geometry and material used during the process. This protocol will also explore the statistical approaches to account for variability in heating parameters, thereby enhancing the reliability of thermal process design. It is intended to guide food processing specialists in establishing safe and efficient thermal processes for a wide range of food products.


**Key words** Heat penetration parameters, Thermal processing, Heating rate index, Lag factor

## 1 Introduction

Thermal processing is a well-established method used in the food industry to ensure product safety, extend shelf life, and achieve commercial sterility in hermetically sealed containers. The effectiveness of such processes depends mainly on the proper estimation of heat penetration parameters, which describe the time–temperature behavior of a food product during sterilization or pasteurization. Accurate determination of these parameters is essential to ensure that the product reaches the required lethality to inactivate pathogenic and spoilage microorganisms, particularly *Clostridium botulinum* in low-acid foods [1, 2].

Heat transfer in foods during retort processing is influenced by several factors, including the physical and rheological properties of the food (e.g., viscosity, density, and particulate content), the

Anderson S. Sant'Ana (ed.), *Heat Resistant Microorganisms in Food Safety and Spoilage*, Methods and Protocols in Food Science, https://doi.org/10.1007/978-1-0716-5280-0_14,

geometry and material of the container (e.g., metal cans, glass jars, pouches, and semi-rigid trays), the mode of heat transfer (conduction, convection, or mixed), and the processing conditions, such as agitation, rotation, and retort medium [3, 4].

In conduction-heated systems (e.g., dense or viscous foods), the cold spot—the last point to reach the target temperature—is typically located at the geometric center of the container. In contrast, for convection-heated foods (e.g., liquids), the cold spot often shifts to a location approximately one-fifth the height from the base [5]. The correct identification and monitoring of this cold spot are essential for generating reliable temperature–time curves, from which critical heat penetration parameters, such as *fH* (heating time factor), *jH* (heating lag factor), and *F0* (sterilization value), are calculated.

Several methods have been developed to model and interpret heat penetration data, including the Ball Formula Method, the Stumbo Method, and the General Method, each with its mathematical assumptions and applications [2, 6, 7]. These models help determine the adequacy of the thermal process and are instrumental in establishing safe and efficient thermal schedules in compliance with regulatory guidelines. Two dominant methods will be addressed in this protocol:

1. Ball Formula Method [6]: A simplified formula-based approach that calculates thermal lethality by assuming log-linear heating and cooling conceptual frameworks with empirical correction factors.
2. General (or Stumbo) Method: A more comprehensive numerical integration method considered more accurate but computationally intensive, often providing shorter processing times—by approximately 15–25%—than the Ball Method, especially in broken heating scenarios [3].

This protocol outlines a scientifically grounded and structured approach for estimating heat penetration parameters in thermally processed foods. It addresses practical steps for thermocouple installation in different container types, data acquisition, cold spot identification, curve analysis, and validation of thermal process adequacy. The procedures described herein are aligned with best practices recommended by international regulatory bodies and thermal processing experts.

## 2 Materials

Consider all the materials and equipment necessary to conduct the heat penetration study (described in Chap. 13) and accurately determine the heating and cooling curves for the specific combination: food × container type × thermal process applied. Also, make sure **Note 1** was addressed.

- Heat processing system (e.g., retort)
- Thermocouples or thermometers with a data logger (to monitor the temperature)
- Samples of food products in appropriate containers (e.g., cans or jars)
- Stopwatch or timer
- Computer with data analysis software (e.g., Excel)
- Sterile tools for handling the food
- Pressure gauge (for retorts that require pressure control)

## 3 Methods

### 3.1 Preparation Phase

1. Place the food product into containers (e.g., cans and jars) that are sealed as per industry standards.
2. Ensure the product and container are suitable for the intended thermal process (e.g., low-acid foods require retort canning).
3. Insert thermocouples at the predetermined cold spot (the slowest heating point) of the product.
4. Seal the containers hermetically to ensure the integrity of the thermocouple and prevent leakage.
5. Place containers in the retort, ensuring uniform placement for reproducibility. Use at least 20 instrumented containers, as described in Chap. 13.
6. Connect the thermocouples to a data logger to continuously record temperature data.

### 3.2 Temperature Measurement and Heating Process

1. Measure the initial temperature ($T_0$) of the food before processing using thermocouples inserted into the coldest part of the product (usually the geometric center of the container).
2. Load the containers into the heat processing system (e.g., retort).
3. Set the system to the desired sterilization temperature ($T_L$) and begin the thermal process.
4. Start the timer to record the product temperature at the beginning of the process ($T_0$).
5. Measure the come-up time ($t_{cut}$): The time necessary for the equipment (retort) to achieve the process temperature ($T_L$).
6. Measure $t_r$: Time during which the retort is maintained at the process temperature ($T_L$).

### 3.3 Data Collection

#### 3.3.1 Monitor Temperature

1. Continuously monitor the temperature at the coldest point in the food (measured by the thermocouples): $T(t)$.
2. Record temperature data at regular intervals (e.g., every 15–30 s).
3. Continue collecting temperature data until the product reaches and holds the desired sterilization temperature ($T_g$).

   OBS: Generally, at the maximum temperature set, the product's cold temperature is about 1 °C lower than at the retort temperature ($T_L$).
4. Once the desired process time or sterilization value is achieved, turn off the heat source.
5. Measure the cooling rate if required. In some protocols, heat penetration testing includes monitoring the cooling phase.
6. Cool the containers rapidly using a controlled cooling process (e.g., water bath or forced air cooling) while continuing to monitor the internal temperature of the food. Record the temperature of the cooling medium ($T_w$).

## 4 Data Analysis Using Ball's Method

1. Choose the microorganism of concern (e.g., *Clostridium botulinum* for low-acid foods) and identify: D-value at the reference temperature (e.g., 121.1 °C); z-value (e.g., 10 °C); and $F_0$-value (e.g., 12D for botulinum cook). When using this method, take observations **Notes 2–4** into consideration.
2. Using the collected data, plot the temperature at the coldest point against time on a graph. The resulting curve should display the temperature rise during heating and the subsequent decrease during cooling, as illustrated in Fig. 1.
3. Determine the initial product temperature ($T_0$) before heating.
4. Plot a semi-logarithmic graph:
   - x-axis (linear): time (min).
   - y-axis (logarithmic): ($T_L$—$T(t)$) vs. time, where: $T_L$ = temperature of the heating medium; $T(t)$ = temperature of the product along the process time.
5. Identify the linear portion of the heating curve (after lag phase).
6. Determine the heating rate factor ($f_H$) using empirical formulas derived from experimental heating and cooling curves, commonly referred to as Ball's method, as shown in Eq. 1:

$$\log\ (T_L - T(t)) = -\frac{t}{f_H} + \log(j_H \cdot (T_L - T_0) \qquad (1)$$

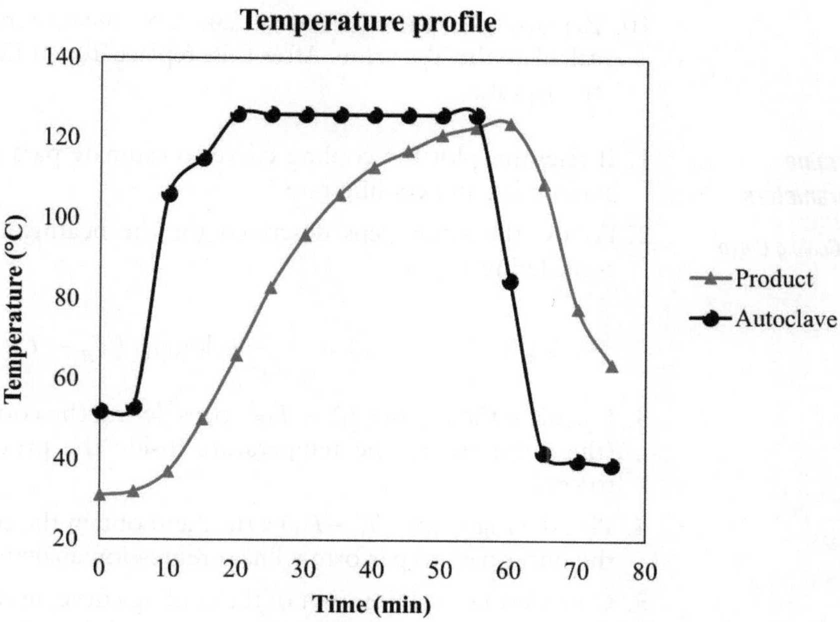


**Fig. 1** Temperature profile recorded during thermal processing in a retort system. The black curve represents the temperature of the heating medium (autoclave chamber), while the gray curve corresponds to the internal temperature at the coldest point of the product

7. Process time ($t_b$) is a calculated process time used to achieve a target lethality ($F_0$ value) in thermal processing. It includes the effective portion of the come-up time ($t_{cut}$), which shows how long the autoclave takes to reach the process temperature. Ball's criterion considers that in 42% of $t_{cut}$, the autoclave is already at the process temperature. So, mathematically: $tb = 0{,}42.t_{cut} + t_r.$

8. After linear regression analysis, the angular coefficient (a) will be related to $f_H$ by the formula: $a = -1/f_H$.. Then, $f_H = -1/a$. It represents the time (in min) to traverse one cycle in this straight-line section.

9. The $j_H$ parameter represents the delay in heating between the container surface and the cold spot. It can be calculated by Eq. 2:

$$j_H = \frac{T_L - T_{p0}}{T_L - T_0} \quad (2)$$

where $T_{P0}$ is the corrected initial temperature in the product, which will be adjusted to the initial corrected time $t_0$', described as 58% of the come-up time, according to Ball's method. So, at time $t_0$', we will have $\log(T_L - T_{P0})$.

10. Replace $t_0$' in the equation obtained by linear regression and calculate the $T_{P0}$ value. After this, replace $T_{P0}$ in Eq. 2 to get the $j_H$ value.

### 4.1 Determine Cooling Parameters

#### 4.1.1 Plot Cooling Curve

1. If relevant, plot the cooling curve to estimate parameters like cooling lag and cooling rate.
2. Follow the same steps described for the heating curve, but considering Eq. 3:

$$\log\ (\ T(t) - T_W) = -\frac{tc}{f_C} + \log(j_C.(T_g - T_W) \qquad (3)$$

3. Calculate the log of $(T_{(t)} - T_W)$, considering the cooling curve (the point where the temperature inside the product starts to cool.
4. Plot the log graph $(T_{(t)}-T_W)$ x time and obtain the equation of the linear part to perform a linear regression analysis.
5. Considering the linear part of the cooling curve, find $f_C$ and $j_C$, knowing that: $f_C = -1/a$, where a = angular coefficient of the linear regression.
6. The $j_C$ parameter represents the measure of the thermal lag of the product at the critical point during cooling. It can be calculated by Eq. 4:

$$j_C = \frac{T_{Pg} - T_W}{T_g - T_W}, \qquad (4)$$

where $T_{Pg}$ is the corrected maximum temperature achieved after the heating phase.

7. Obtain the value of $T_{Pg}$ using the linear equation of the plot, considering the point of intersection, where the initial time is 0. Replacing 0 in the equation, we obtain the linear coefficient that is correlated with log $(T_{Pg} - T_W)$.

### 4.2 Practical Application

1. Once the heat penetration parameters ($f_H$, $j_H$, $f_C$, $j_C$, $T_0$, $T_g$, $T_{(t)}$), as well as the heat processing parameters ($t_{cut}$, $t_r$, $t_b$, $T_L$, $T_W$), are known, it is possible to determine the process lethality values ($F_P$) equivalent to a given reference temperature ($F_0$). In this case, the z value of the microorganism must be known.
2. There are two situations to apply Ball's Mathematical Method:
   (a) If $F_P$ is known, $t_b$ can be calculated.
   (b) If $t_b$ is known, $F_P$ can be calculated.

#### 4.2.1 Mathematical Method

All mathematical methods are based on semi-logarithmic heating curves, as described in Eq. 1, with some considerations: the sterilization value during the heating phase ($U_h$)—expressed as the equivalent time at a reference temperature $T_L$—is a function of the heating rate index ($f_h$) and the temperature differential ($T_L$—$T_g$), where $T_L$ is the retort temperature. $T_g$ is the final product temperature at the end of heating. So, at the end of heating, Eq. 1 is rewritten as:

$$\log\left(T_L - T_g\right) = -\frac{t_b}{f_H} + \log(j_H.(T_L - T_0) \quad (5)$$

It can also be represented by Eq. 6:

$$\log(g) = -\frac{t_b}{f_H} + \log(j_H.(T_L - T_0) \quad (6)$$

So, when the thermal process conditions—such as process time ($tb$), retort temperature ($T_L$), initial product temperature ($T_0$), heating rate index ($f_h$), and heating lag factor ($j_h$)—are known, the value of $g$ (temperature deficit factor) can be directly calculated. Subsequently, the equivalent lethality ($U$) and the process lethality value ($F_P$) can be determined, allowing for a complete evaluation and validation of the thermal process.

When the retort temperature ($T_L$), initial product temperature ($T_0$), heating rate index ($f_h$), and heating lag factor ($j_h$) are known, along with the desired process lethality value ($F_P$), the equivalent lethality ($U$) can be calculated. From this, the ratio $R = fh/U$ can be determined, enabling the calculation of log($g$) and, subsequently, the required process time ($t_b$), following the process design mode of the Ball method. Both situations can be better visualized in Fig. A1 and A2. For this analysis, it is necessary to have access to the log ($g$) × R table, calculated by Stumbo [2]. These tables allowed processors to avoid complex differential equations and design thermal schedules directly from the known or assumed values of $f_H$, $T_L$, $T_0$, and $F_P$

## 5 Data Analysis Using the General Method

1. For the application of the general method, experimentally determined time–temperature data in the cold spot are necessary.
2. Calculate the lethal rate for each time interval, which compares the microorganism's killing effect at a measured temperature to that at 1 min at a reference temperature. This is a crucial step in determining the F-value.
3. Multiply the lethal rate by the time to obtain $F_P$ (sterilization value estimated from heat penetration data). It reflects the lethality of a thermal process in terms of equivalent time at a reference temperature.

**Step 1:** Calculate log (TL-T(t)) for each point:

| Tempo (min) | Temperature (°F) | log (TL-T(t)) | |
|---|---|---|---|
| 0 | 45.0 | =LOG(250-45)= | 2.31 |
| 2 | 48.0 | =LOG(250-48)= | 2.31 |
| 4 | 54.0 | =LOG(250-54)= | 2.29 |
| 6 | 64.0 | =LOG(250-64)= | 2.27 |
| 8 | 79.0 | =LOG(250-79)= | 2.23 |
| 10 | 97.0 | =LOG(250-97)= | 2.18 |
| 12 | 119.0 | =LOG(250-119)= | 2.12 |
| 14 | 145.0 | =LOG(250-145)= | 2.02 |
| 16 (Autoclave at TR) | 174.0 | =LOG(250-174)= | 1.88 |
| 18 | 203.0 | =LOG(250-203)= | 1.67 |
| 20 | 220.0 | =LOG(250-220)= | 1.48 |
| 22 | 232.0 | =LOG(250-232)= | 1.26 |
| 24 | 238.5 | =LOG(250-238.5)= | 1.06 |
| 26 | 242.0 | =LOG(250-242)= | 0.90 |
| 28 | 245.0 | =LOG(250-245)= | 0.70 |
| 30 | 246.5 | =LOG(250-246.5)= | 0.54 |
| 32 | 248.0 | =LOG(250-248)= | 0.30 |
| 34 | 248.5 | =LOG(250-248.5)= | 0.18 |
| 36 | 249.0 | =LOG(250-249)= | 0.00 |
| 38 | 249.5 | =LOG(250-249.5)= | -0.30 |
| 40 (Cooling phase starts) | 250.0 | =LOG(250-250)= | #NUM! |
| 42 | 250.0 | =LOG(250-250)= | #NUM! |
| 44 | 238.0 | =LOG(250-238)= | 1.08 |
| 46 | 212.0 | =LOG(250-212)= | 1.58 |
| 48 | 180.0 | =LOG(250-180)= | 1.85 |
| 50 | 155.0 | =LOG(250-155)= | 1.98 |

**Step 2:** Plot the log (TL-T(t)) versus time (heating curve), highlighted in red:

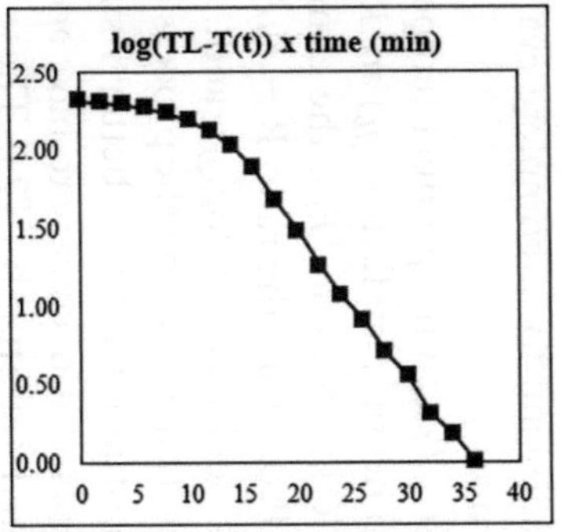


**Step 3:** Set the linear portion of the heating curve (after lag phase) and determine the linear regression analysis

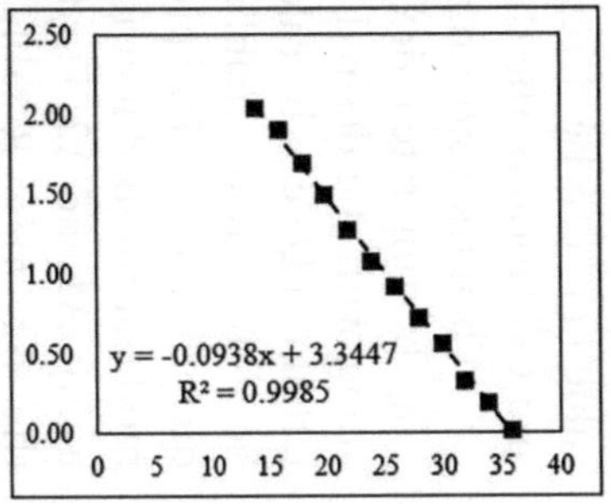


**Step 4:** Determine the heating rate factor (fH) using empirical formulas derived from experimental heating curve:

y = ax + b ⟶ y = -0.0938x + 3.3447

$$\log\left(T_L - T(t)\right) = -\frac{t}{f_H} + \log\left(j_H \cdot (T_L - T_0)\right)$$

$$-\frac{1}{f_H} = -0.0938 \longrightarrow f_H = \frac{1}{0.0938} = 10.66 \text{ minutes}$$

**Step 5:** Calculate the initial corrected time t0', described as 58% of the come up time, according to Ball's method.

t0' = 0.58 x 16 = 9.28 minutes

**Step 6:** Replace it in the same equation to obtain the term (TL - TP0):

y = -0.0938x + 3.3447 ⟶ log(TL-TP0)= -0.0938*9,28t + 3.3781

⟶ log(TL-TP0) = 2.47

⟶ TL - TP0 = 10^(2.47) = 298.01

**Step 7:** Calculate the jH parameter by the formula:

$$j_H = \frac{T_L - T_{p0}}{T_L - T_0} = \frac{298.01}{250 - 45} = 1.45$$

**Fig. A1** Step-by-step application of the Ball method for determining the heating rate factor ( $f_h$) and related thermal process parameters based on experimental data

**Step 1:** Calculate log (T(t)-Tw) for each point:

| Time (min) | Tempo (min) | Temperature (°F) | log (T(t)-Tw) | |
|---|---|---|---|---|
| | 0 | 45.0 | | |
| | 2 | 48.0 | | |
| | 4 | 54.0 | | |
| | 6 | 64.0 | | |
| | 8 | 79.0 | | |
| | 10 | 97.0 | | |
| | 12 | 119.0 | | |
| | 14 | 145.0 | | |
| | 16 (Autoclave at TR) | 174.0 | | |
| | 18 | 203.0 | | |
| | 20 | 220.0 | | |
| | 22 | 232.0 | | |
| | 24 | 238.5 | | |
| | 26 | 242.0 | | |
| | 28 | 245.0 | | |
| | 30 | 246.5 | | |
| | 32 | 248.0 | | |
| | 34 | 248.5 | | |
| | 36 | 249.0 | | |
| | 38 | 249.5 | | |
| 0.00 | 40 (Cooling phase starts) | 250.0 | =LOG(70-250)= | 2.32 |
| 2.00 | 42 | 250.0 | =LOG(70-250)= | 2.32 |
| 4.00 | 44 | 238.0 | =LOG(70-238)= | 2.30 |
| 6.00 | 46 | 212.0 | =LOG(70-212)= | 2.24 |
| 8.00 | 48 | 180.0 | =LOG(70-180)= | 2.15 |
| 10.00 | 50 | 155.0 | =LOG(70-155)= | 2.06 |

**Step 2:** Plot the log (T(t)-Tw) versus time (cooling curve), highlighted in blue:

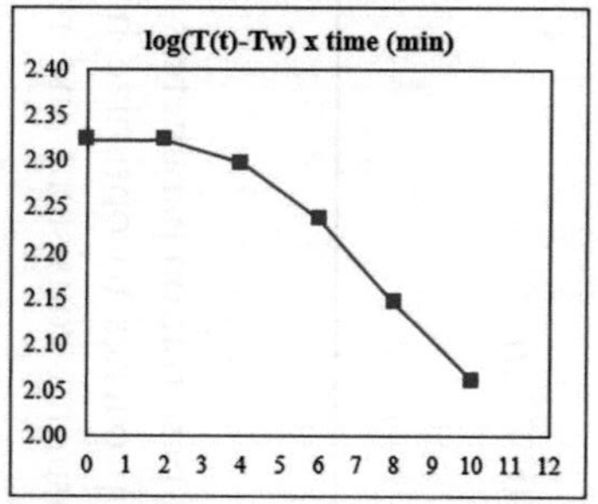


**Step 3:** Set the linear portion of the cooling curve (after lag phase) and determine the linear regression analysis.

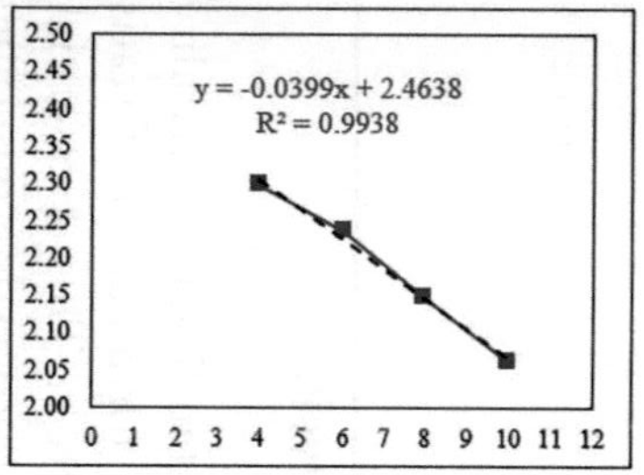


**Step 4:** Determine the cooling rate factor (fC) using empirical formulas derived from experimental cooling curve:

$y = ax + b \longrightarrow y = -0.0399x + 2.4638$

$$\boxed{log\left(T(t) - T_w\right) = -\frac{tc}{f_c} + log\left(j_c \cdot (T_g - T_w)\right)}$$

$$-\frac{1}{fc} = -0.0399 \longrightarrow fc = \frac{1}{0.0399} = 25.06 \text{ minutes}$$

**Step 5:** Replacing 0 in the equation, we will have the linear coefficient that is correlated to log (TPg - TW)

y = -0.0399x + 2.4638 ⟶log(TPg - Tw) = -0.0399 * 0 + 2.4638

⟶ TPg - Tw = 10^(2.4638)

⟶ TPg - Tw = 290.94

**Step 7:** Calculate the jH parameter by the formula:

$$j_c = \frac{T_{Pg} - T_W}{T_g - T_W} = \frac{290.94}{250 - 70} = 1.62$$

**Fig. A2** Step-by-step application of the Ball method for determining the cooling rate factor ($f_C$) and related thermal process parameters based on experimental data

4. Integrate the lethal rates: Use the trapezoidal method to integrate the lethal rates over the measured times and temperatures, converting the time–temperature data into F-values that represent the process lethality. This calculation is typically embedded within data logger software, enabling real-time monitoring or post-processing evaluation. It is possible to do it in an Excel table using Eq. 7.
5. Sum up all calculated F-values to have the cumulative *Fp* delivered by the process.
6. Check the example performed in Fig. A3:

$$FP = \int_{ti}^{tf} L.dt \tag{7}$$

where $L = 10^{\frac{T-250}{z.}}$

$$\mathrm{FP} = .\int_{ti}^{tf} 10^{\frac{T-250}{z}}.dt$$

## 6 Statistical Variability Analysis

Understanding the variability in heat penetration parameters allows for the formulation of statistical approaches to optimize heating parameters. This is essential for ensuring consistent and reliable thermal processing outcomes [8].

1. To obtain a reliable picture of the statistical variability of heating parameters, heat penetration tests with 30 to 40 test containers are recommended. Using more samples beyond this range may not significantly affect the estimated heating characteristic.
2. Calculate the mean, variance, and coefficients of variation (CV) for the parameters. This helps in understanding the extent of variability and comparing it with literature data.
3. Analyze run-to-run variability using statistical tests such as the F-test and the Tukey test to identify significant differences between runs.
4. Reported standard deviations for fh range from 7% to 25% of its value. The CV for the heating rate index (*fh* value) can range from 2.4% to 11.4% for different products. Consider the curve that represents the worst-case scenario.

By following these steps, we can accurately estimate the heat penetration parameters of foods to ensure they meet the necessary safety and quality standards during thermal processing, as well as product safety requirements.

The following data represent the temperature values of a food in the slowest heating point over the course of a thermal process carried out at 250 F. Calculate the $F_0$ delivered by this process.

**Step 1:** Consider *C. botulinum* as the target microorganism and the parameters: $D_0 = 0{,}21$ min; $Z = 18$ °F; $T_{ref} = 250$ °F.

**Step 2:** Calculate the letality of each time point by using the formula: $L = 10^{\frac{T-T_{ref}}{z}}$

**Step 3:** Calculate the time intervals along the process (Δt)

**Step 4:** Calculate the sterilization value ($\Delta F_P$) delivered by each process time interval: $\Delta Fp = [(L_n + L_{n-1})/2] * \Delta t$, considering the trapezoidal method.

| Tempo (min) | Temperature (°F) | L | Δt | ΔFp = Δt . ((L n+Ln-1)/2) |
|---|---|---|---|---|
| 0 | 142 | L= 10 ^ ((142 - 250)/18) = 0,000 | = 0 | |
| 5 | 142 | 0,000 | = 5 - 0 = 5 | 0,00 |
| 10 | 142 | 0,000 | = 10 - 5 = 5 | 0,00 |
| 15 | 142 | 0,000 | = 15 – 10 = 5 | 0,00 |
| 20 | 165 | 0,000 | = 20 – 15 = 5 | 0,00 |
| 25 | 187 | 0,000 | = 25 – 20 = 5 | 0,00 |
| 30 | 203 | 0,002 | = 30 - 25 = 5 | 0,01 |
| 35 | 215 | 0,011 | = 35 – 30 = 5 | 0,03 |
| 40 | 226 | 0,046 | = 40 - 35 = 5 | = 5 . ((0,046 + 0,011)/2) = 0,14 |
| 45 | 231,4 | 0,093 | = 45 – 40 = 5 | 0,35 |
| 50 | 236,5 | 0,178 | = 50 – 45 = 5 | 0,68 |
| 55 | 240 | 0,278 | = 55 – 50 = 5 | 1,14 |
| 60 | 243 | 0,408 | = 60 – 55 = 5 | 1,72 |
| 65 | 244 | 0,464 | = 65 – 60 = 5 | 2,18 |
| 70 | 247 | 0,681 | = 70 – 65 = 5 | 2,86 |
| 75 | 248,3 | 0,805 | = 75 – 70 = 5 | 3,71 |
| 80 | 249,3 | 0,914 | = 80 – 75 = 5 | 4,30 |
| 85 | 249 | 0,880 | = 85 – 80 = 5 | 4,49 |
| 90 | 247,2 | 0,699 | = 90 – 85 = 5 | 3,95 |
| 95 | 225,5 | 0,044 | = 95 – 90 = 5 | 1,86 |
| 100 | 177 | 0,000 | = 100 – 95 = 5 | 0,11 |
| 105 | 155 | 0,000 | = 105 – 100 = 5 | 0,00 |

**Step 5:** Sum all calculated ΔFp to obtain the $F_P$.

$F_P = 0{,}01 + 0{,}03 + 0{,}14 + 0{,}35 + 0{,}68 + 1{,}14 + 1{,}72 + 2{,}18 + 2{,}86 + 3{,}71 + 4{,}30 + 4{,}49 + 3{,}95 + 1{,}86 + 0{,}11 + 0{,}0 = 27{,}52$ min

**Conclusion**

The thermal process delivered a lethality equivalent to 27.52 min at an isothermal process performed at 250 °F.

**Fig. A3** Step-by-step protocol to calculate the sterility value ($F_P$) of a thermal process by the general method

## 7 Notes

1. Ensure the retort or heating system is calibrated and functioning correctly to avoid errors in heat penetration estimation.
2. The Ball method is generally applied to low-acid foods in retort systems [3].
3. The size and shape of the container will affect heat penetration and should be taken into account during analysis [Refer chapter 13 of this book].
4. Make sure all temperatures are set to °F before applying Ball's mathematical method [3].

## Acknowledgments

L. P. Margalho and E.P.E. Sitoe acknowledge the financial support of the "Fundação de Amparo à Pesquisa do Estado de São Paulo" (FAPESP): Grant #2020/14631-6 and #2024/20257-0. This study was financed in part by the "Coordenação de Aperfeiçoamento de Pessoal de Nível Superior" (CAPES)—Finance Code 001.

## Annex

### References

1. Ziaiifar AM, Nedamani AR (2023) Thermal food process calculations. In: Thermal processing of food products by steam and hot water. Woodhead Publishing, pp 27–66
2. Stumbo CR (2013) Thermobacteriology in food processing. Elsevier
3. Holdsworth SD, Simpson R, Barbosa-Cánovas GV (2008) Thermal processing of packaged foods, vol 284. Springer, New York
4. Fellows PJ (2022) Food processing technology: principles and practice. Woodhead Publishing
5. Simpson R, Ramírez C (2020) Principles of thermal processing of packaged foods. Princ. Therm. Process. Packag, Foods
6. Ball CO (1923) Thermal process time for canned food (vol. 7, no. 37). National Research Council of the National Academy of Sciences
7. Toledo RT, Toledo RT (2007) Thermal process calculations. In: Fundamentals of food process engineering, pp 301–378
8. Smout C, Loey AV, Hendrickx M, Beirlant J (2000) Statistical variability of heat penetration parameters in relation to process design. J Food Sci 65(4):685–693

# Chapter 15

# Preparing Biological Indicators for Validation of Thermal Processes

**Wilma Custódio Fumo and Marcelo Felipe Silva Estácio Santana**


## Abstract

Thermal validation processes require the use of biological indicators (BIs) to assess the microbial reductions achieved by heat treatments, thereby ensuring food safaty and product quality. Surrogates microorganisms are often used as Bls because they allow validation studies to be conducted without introducing pathogens into processing environments. The selection of surrogate microorganisms with thermal resistence comparable to that of target phatogens is fundamental to ensuring study robustenes and reproducibility. This chapter presents preparation and inoculation methods for BIs used in thermal validation studies, using *Enterococcus faecium* NRRL B-2354, as a surrogate of *Salmonella enterica*, as an example of BI. Direct inoculation method for high-moisture foods. and dry inoculation approaches using carrier agents(such talc, sand, etc.) for low-moisture food a described, together with critical microbiological and operational parametes influencing process effectivenes.


**Key words** Water activity, *Enterococcus faecium* NRRL B-2354, Microorganism

## 1 Introduction

Validation of thermal processes is an important study which demonstrates the impact of the heat processes in reducing the microbial population in food and beverages. Several research and reviews on validation and inoculation- studies have been published over the years given the importance on the context of food safety [1].

As described in Chap. 16 (Conducting a Validation Study of Thermal Processing with Biological Indicators), the main step of validation studies is the intentional inoculation of food samples for thermal process with specific microorganisms, known as biological indicators (BI). Most validation studies use surrogates of the target microorganisms as BIs, avoiding the introduction of pathogens in the processing line [1]. Table 1 contains some surrogates of pathogens which may be used as BIs in thermal validation

Anderson S. Sant'Ana (ed.), *Heat Resistant Microorganisms in Food Safety and Spoilage*, Methods and Protocols in Food Science, https://doi.org/10.1007/978-1-0716-5280-0_15,

**Table 1**
**Surrogates microorganisms used as biological indicators in thermal validation studies**

| Target microorganism | Surrogates | References |
|---|---|---|
| *Clostridium botulinum* | *Clostridium sporogenes* PA 3679 | [2] |
| *Salmonella enterica* | *Enterococcus faecium* NRRL B-2354<br>*Pediococcus acidilactici* ATCC 8042 | [3]<br>[4] |
| *Listeria monocytogenes* | *Listeria innocua* ATCC 51724 | [5] |

studies. However, the application depends on the food matrix and the kind of thermal process.

In this chapter, we describe the preparation of BI to be used in microbiological challenge tests of validation studies of thermal processes for both types of food (low- and high-moisture foods), using *Enterococcus faecium* NRRL B-2354 as a example of BI.

## 2 Materials

### 2.1 Strains

In this chapter, the materials described below are used in the cultivation of *Enterococcus faecium* NRRL B-2354, applied as the surrogate of *Salmonella enterica* in thermal processes, especially in low-moisture foods.

### 2.2 Media and Diluents

1. *Trypticase Soy Broth (TSB)*: 17 g casein peptone (enzymatic digestion), 3 g soy peptone (enzymatic digestion), 5 g sodium chloride (NaCl), 2.5 g dipotassium phosphate (K2HPO4), 2.5 g dextrose, and 1 L of purified water. Adjust the final pH to 7.3 ± 0.2 using 1 N sodium hydroxide (NaOH) solution while stirring. Autoclave at 121 °C for 15 min.
2. *Trypticase Soy Agar (TSA):* 15 g Casein peptone (enzymatic digestion), 3 g soy peptone (enzymatic digestion), 5 g sodium chloride (NaCl), and 1 L purified water. Adjust the final pH to 7.3 ± 0.2 with 1 N NaOH solution while stirring. Autoclave at 121 °C for 15 min.
3. *Xylose Lysine Deoxycholate Agar (XLD):* Suspend 52.9 g of the medium in 1 L of purified water; heat the medium to 90 °C while stirring and stop when the medium is completely dissolved. Do not autoclave.
4. *Peptonized saline water (H2O sp)*: 8.5 g Sodium chloride (NaCl) and 1 g peptone in 1 L of purified water.
5. *Peptone water*: 1 g peptone in 1 L of purified water.
6. *Saline water*: 8.5 g NaCl in 1 L of purified water.

**2.3 Carrier Agents**

1. Talc.
2. Sand.
3. Calcium carbonate and silicon dioxide ($SiO_2$).

# 3 Methods

Figure 1 shows the steps to obtain BI inoculum to use in thermal validation studies.

**3.1 Inoculum Preparation**

The inoculum (*E. faecium* NRRL B-2354) is prepared as follows, as instructed in Chap. 5 (Methods for Preparation of Suspensions of Bacterial Vegetative Cells). If a suspension of spores or fungi/yeast is used as BI, refer Chaps. 6 (Methods for Preparation of Suspensions of Bacterial Spores) and 7 (Methods for Preparation of Suspensions of Filamentous Fungi and Yeasts).

**3.2 Preparation of Carrier Agents**

1. Sterilize the carrier at 140 ° C for 4 h in an oven [6], or for sand at 121 °C for 15 min [7, 8], for decontamination (if necessary, if not sterile).
2. Transfer the previously standardized microganism suspension (Chap. 5) to the carrier agent in a ratio of 1:1 for carbonate and silica, 1:2 for talc and sand in a sterile container, homogenize

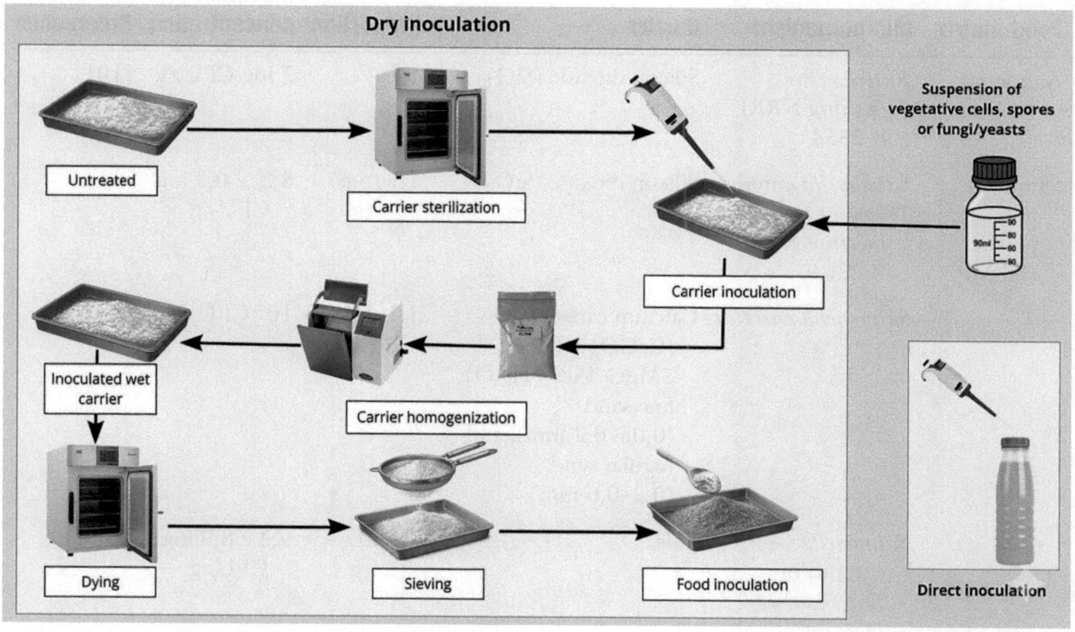


**Fig. 1** Simplified scheme for the preparation of dry and wet inoculum for validation studies of thermal processes. (The authors)

manually for 3 to 5 min or using a stomacher at 260 rpm for 3 min.

3. Transfer to a sterile metal or glass tray, spread in a thin layer (3 to 5 mm), and cover with an equally sterile aluminum foil.
4. Take the oven preheated to 35 °C/24 h; if necessary, let it dry at room temperature for another 24 h until the material reaches an $a_w$ value of 0.44-0.46 or equal to the food matrix under test.
5. After drying, sift the vehicle with a sieve of 74–100 μm to break up any agglomerate (in the case of calcium carbonate, it may be necessary to break up the lumps with a mortal and pestle).
6. Transfer to a sterile container and store at 4 °C for 2 days.

### 3.3 Inoculation in Low-Moisture Foods

1. *Inoculation:* After preparing the dry inoculum, add the inoculated carrier to the food matrix in a sterile container (*see* **Note 1**) and homogenize manually for 5 min.
2. *Checking the cell concentration in food:* Check the first concentration of cells using the method described in Table 2. The concentration of cells in the food matrix must be at least

**Table 2**
**Studies on dry and low inoculation for thermal validation studies**

| Food matrix | Microorganism | Carrier | Proportion | Inoculum concentration | References |
|---|---|---|---|---|---|
| Almonds | *Enterococcus faecium* NRRL B-2354 | Silicon dioxide ($SiO_2$) | 1:1 | 7 log CFU/g | [10] |
| Skimmed milk powder (NFMP) | *Salmonella* enterica *Enterococcus faecium* NRRL B-2354 | Silicon dioxide ($SiO_2$) | 1:1 (p:p) | 8,2 ± 0,2 log CFU/g | [11] |
| — | *Salmonella enterica* | Calcium carbonate ($CaCO_3$); Talco $3MgO{\cdot}4SiO_2.H_2O$); fine sand (0,05–0,2 mm); and regular sand (0,2–0,6 mm) | 1:1; 1:2; 1: 2 (p:p) | $10^8$ CFU/g | [8] |
| — | *Salmonella* enterica or *Enterococcus faecium* | Talc | 18 mL: 25 g | 9,2 e 8,9 log CFU/g | [6] |
| Soya meal | *Salmonella enterica* | Sand | 1:2 (mL: g) | $10^9$ CFU/g | [7] |

2-log cycles higher than the aimed log-reduction designed for the process for the target BI.

3. *Checking changes in water activity ($a_w$):* Measure $a_w$ before and after inoculation, both of the vehicle and of the feed matrix under test, so there is no change.

### 3.4 Inoculation in High-Moisture Foods

1. *Inoculation:* Add the liquid inoculum directly to the food using a pipette (200 μL, 1000 μL, or higher) with a sterilized tip. The inoculum volume should not exceed 1% of the sample (e.g., in 100 g of food, the volume of inoculum must be up to 1000 μL) [9]. The inoculation may be made on-site, to facilitate and avoid changes during shipping from the laboratory to the test facility [1].
2. *Checking the cell concentration in food:* Check the initial concentration of cells using the method described in Subheading 3.5. The concentration of cells in the food matrix must be at least 2-log higher than the aimed log reduction designed to the process for the target microorganism [1].
3. *Checking the changes in water activity ($a_w$) and pH:* Before and after the inoculation, measure $a_w$ and pH of the samples to verify if the inoculation caused undesirable changes. In case of changes, add a humectant and an acidulant in the buffer to adjust $a_w$ and pH, respectively [9].

### 3.5 Enumeration of Cells

1. After storage, weigh 25 g of the vehicle in a sterile bag or container and mix with 225 mL of H2Osp water or 0.1% of peptone water.
2. Homogenize the mixture with the help of a stomacher at normal speed for 30 s or shake by hand.
3. Perform serial dilutions in test tubes containing NaCl or peptone water ($10^{-2}$; $10^{-4}$; $10^{-6}$) and consider the homogenized content as the first dilution.
4. Transfer/sow a 1 mL aliquot of the diluted sample in Petri dishes containing TSA agar or XLD agar.
5. Incubate the plates at 35 ° C for 24/48 h in the case of TSA or incubate at 37 ° C for 24 h in the case of XLD.
6. After incubation, count plates that have between 25 and 250 colonies and calculate the results in CFU/g.

## 4 Note

1. The proportions of inoculation from the vehicle to the matrix depend on the physicochemical characteristics of the same [6, 11–13].

## References


1. Ceylan E, Amezquita A, Anderson N, Betts R, Blayo L, Garces-Vega F et al (2021) Guidance on validation of lethal control measures for foodborne pathogens in foods. Compr Rev Food Sci Food Saf 20(3):2825–2881
2. Brown JL, Tran-Dinh N, Chapman B (2012) *Clostridium sporogenes* PA 3679 and its uses in the derivation of thermal processing schedules for low-acid shelf-stable foods and as a research model for proteolytic clostridium botulinum. J Food Prot [Internet] 75(4):779–792. Available from: https://doi.org/10.4315/0362-028X.JFP-11-391
3. Jeong S, Marks BP, Ryser ET (2011) Quantifying the performance of *Pediococcus sp.* (NRRL B-2354: enterococcus faecium) as a nonpathogenic surrogate for *Salmonella Enteritidis* PT30 during moist-air convection heating of almonds. J Food Prot [Internet] 74(4): 603–609. Available from: https://doi.org/10.4315/0362-028X.JFP-10-416
4. Deen B, Diez-Gonzalez F (2019) Assessment of *Pediococcus acidilactici* ATCC 8042 as potential *Salmonella* surrogate for thermal treatments of toasted oats cereal and peanut butter. Food Microbiol [Internet] 83 (May):187–192. Available from: https://doi.org/10.1016/j.fm.2019.05.015
5. Shi Y, Tang J, Yue T, Rasco B, Wang S (2015) Pasteurizing cold smoked Salmon (Oncorhynchus nerka): thermal inactivation kinetics of *Listeria monocytogenes* and *Listeria innocua*. J Aquat Food Prod Technol [Internet] 24(7): 712–722. Available from: https://doi.org/10.1080/10498850.2013.808303
6. Enache E, Kataoka A, Glenn Black D, Napier CD, Podolak R, Hayman MM (2015) Development of a dry inoculation method for thermal challenge studies in low-moisture foods by using talc as a carrier for *Salmonella* and a surrogate (Enterococcus faecium). J Food Prot [Internet] 78(6):1106–1112. Available from: https://doi.org/10.4315/0362-028X.JFP-14-396
7. Norberto AP, Alvarenga VO, Hungaro HM, AS S'A (2022) Desiccation resistance of a large set of *Salmonella enterica* strains and survival on dry- and wet-inoculated soybean meal through storage. LWT 158
8. Furtado MM, Silva BS, Faviero C, Alvarenga VO, Sant'Ana AS (2020) Impact of carrier agents and temperature during storage of dry inocula of *Salmonella enterica*: a contribution to the validation of low water activity foods processing interventions. LWT [Internet] 131:109705. Available from: https://doi.org/10.1016/j.lwt.2020.109705
9. NACMCF Executive (2010) Parameters for determining inoculated pack/challenge study protocols. J Food Prot [Internet]. 73(1): 140–202. Available from: https://doi.org/10.4315/0362-028X-73.1.140
10. Ly JU (2014) Guidelines for using *Enterococcus faecium* NRRL B-2354 as a surrogate microorganism in almond process validation. Almond Board Calif:1–12
11. Liu S, Xu J, Xie L, Zhu MJ, Tang J (2019) Dry inoculation methods for nonfat milk powder. J Dairy Sci [Internet] 102(1):77–86. Available from: https://doi.org/10.3168/jds.2018-14478
12. Blessington T, Theofel CG, Harris LJ (2013) A dry-inoculation method for nut kernels. Food Microbiol [Internet] 33(2):292–297. Available from: https://doi.org/10.1016/j.fm.2012.09.009
13. Xu J, Liu S, Song J, Tang J, Zhu MJ, Gray P et al (2018) Dry-inoculation method for thermal inactivation studies in wheat flour using freeze-dried *Enterococcus faecium* NRRL B-2354. LWT [Internet] 89:10–17. Available from: https://doi.org/10.1016/j.lwt.2017.10.006
14. Ahn J, Balasubramanian VM (2007) Effects of inoculum level and pressure pulse on the inactivation of *Clostridium sporogenes* spores by pressure-assisted thermal processing. J Microbiol Biotechnol 17(4):616–623

# Chapter 16

# Conducting a Validation Study of Thermal Processing with Biological Indicators

**Marcelo Felipe Silva Estácio Santana**


## Abstract

Validation study of thermal processing with biological indicators is an important test in the context of food safety. The process involves at least four key-steps: 1-sample preparation, 2-thermal processing, 3-microbiological analysis, and 4-data interpretation and documentation. This chapter outlines the procedures to validate a designed thermal processing for both high- and low-moisture food.


**Key words** Integrated lethality, Decimal reduction number, Food safety, *Enterococcus faecium* NRRL B-2354, Hazard Analysis and Critical Control Points

## 1 Introduction

As mentioned in Chap. 15 (Preparing Biological Indicators for Validation of Thermal Processes), the validation study of thermal processing is an indispensable constituent of Hazard Analysis and Critical Control Points (HACCP), since it allows the evaluation of the real fate of a target microorganism during thermal processing.

Generally, a thermal validation study has four steps (Fig. 1). The first step involves preparing and inoculating the food samples with the biological indicator (BI). The second step consists in conducting the designed thermal process. After, the samples are submitted for microbiological analysis, which represents the third and most critical step of the validation process, as it provides the most important data. Lastly, all collected data are analyzed to validate the thermal processing and documented for posterior reanalysis to identify a possible revalidation.

This chapter presents a protocol for conducting validation studies of thermal processing in both low- and high-moisture foods using BIs, specifically *Enterococcus faecium* NRRL B-2354, a recognized surrogate for *Salmonella enterica*.

Anderson S. Sant'Ana (ed.), *Heat Resistant Microorganisms in Food Safety and Spoilage*, Methods and Protocols in Food Science, https://doi.org/10.1007/978-1-0716-5280-0_16,

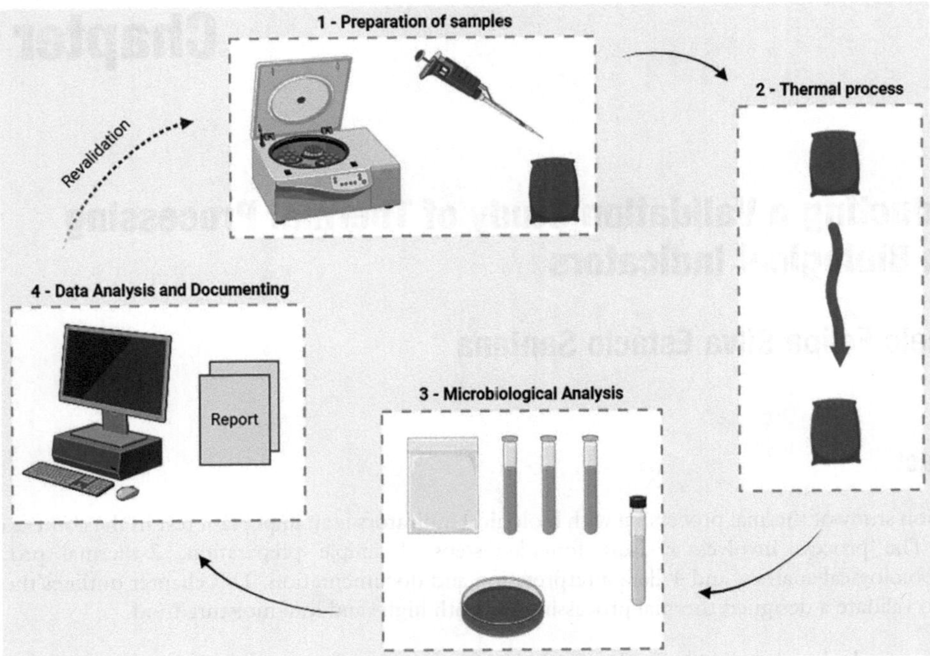


**Fig. 1** General steps of a thermal validation study. (The author)

## 2 Materials

### 2.1 Preparation of Samples

The preparation and formula for the materials can be found in Chap. 15.

### 2.2 Thermal Process

1. *Calibrated thermocouple*: To know how to calibrate thermocouples, refer Chap. 12 (*Calibration of Thermocouples for Temperature Measurement During Thermal Processes*).

### 2.3 Microbiological Analysis

Most of the required materials are listed in Chap. 15. All materials should be prepared using purified water obtained through a reverse osmosis filtration system.

1. *Brain Heart Infusion (BHI)*: Brain heart infusion solids (17.5 g), tryptose (10 g), glucose (2.0 g), sodium chloride (5.0 g), disodium hydrogen phosphate (2.5 g), and water (1 L). Prepare the suspension by mixing all components, dispense into tubes, and autoclave at 121 °C for 15 min. Store the plates at 4 °C until use.
2. *m-Enterococcus agar*: Enzymatic digest of casein (15.0 g), enzymatic digest of soybean meal (5.0 g), yeast extract (5.0 g), dextrose (2.0 g), dipotassium phosphate (4.0 g), sodium azide (0.4 g), 2,3,5-triphenyl tetrazolium chloride

(0.1 g), agar (10.0 g), and water (1 L). Prepare the suspension by mixing all materials, heat with agitation, and boil for 1 min using a microwave (do not autoclave). Cool to 50 °C and fill the plates. Wait for agar solidification with the plate open inside of the laminar flow cabinet. Store the plates at 4 °C until use.

## 3 Methods

### 3.1 Preparation of Samples

1. *Preparation of the bacterial suspension:* To see how to prepare suspensions of bacterial vegetative cells and spores, refer Chaps. 6 and 7 (*Methods for Preparation of Suspensions of Bacterial Vegetative Cells* and *Methods for Preparation of Suspensions of Bacterial Spores*, respectively). For fungi/yeast BIs, refer Chap. 8 (*Methods for Preparation of Suspensions of Filamentous Fungi and Yeasts*).
2. *Inoculation of the samples:* The inoculation procedure should be conducted as described in Chap. 15. Prepare the inoculum to achieve a concentration of 7–8 log CFU/g or mL in food or 2-log above the targeted reduction desired (*see* **Note 1**; [1, 2]).
3. *Preparation of packed samples:* For low-moisture and solid food, add 50 g of the inoculated samples in a suitable container that can be embedded in the processing line [2]. Special attention must be given to the type of container: Choose a material with good thermal conductivity and ensure that the container is securely sealed to prevent it from opening or leaking during the process, without changing the atmosphere inside the pack (*see* **Note 2**). For high-moisture, liquid, and particulate food (beverages, soups, sauces, etc.) that is conditioned in bottles/packs before the thermal process, the inoculated food can be added in its container to simulate a real scenario.
4. *Replication and control samples:* For each time interval, add two or more inoculated samples embedded in the process and the same quantity of control (inoculated but untreated sample [1]. Conduct at least three replicates of the validation process (*see* **Note 3**).

### 3.2 Thermal Process

1. *Process and sample temperature:* Before initiating the validation and the inoculation procedure, for the noncontinuous process, make a map of the temperature of the processing line, finding all the cold spots in the process [2]. Measure the temperature using calibrated thermocouples (*see* **Note 4**) distributed throughout the whole line where the food pack will be embedded. For a continuous process, the treatment should be conducted at the set temperature of the divert valve, as the product will be rejected by the valve if its temperature falls below this threshold [3].

2. *Distribution of the samples:* For conveyor systems, embed all samples within the product bed, ensuring they are positioned among the regular product units. In nonconveyor lines, place the samples along the chamber, attempting to include samples in all the cold spots to evaluate a worst-case scenario [2].
3. *Collecting samples*: Conduct the thermal process and collect all the packed samples before (control) and after the process (treated samples) in the batch. Conduct the thermal process and collect all the packed samples before and after the process (treated samples). In continuous processing, collect enough samples before and after processing, some for immediate microbiological analysis and others to be stored as counterproof. Analyze all samples as described in Subheading 3.3.

### 3.3 Microbiological Analysis

1. *Enumeration:* Conduct the analysis and count the colonies as described in Chap. 15 (*Preparing Biological Indicators for Validation of Thermal Processes*). After mixing the sample with the diluent, store the bags containing all the homogeneous solutions at 4 °C until plate counts are performed.
2. *Enrichment:* In cases where microbial counts are lower than the limit of detection, enrichment is required to confirm the presence or absence of viable organisms. Add 25 mL of the homogeneous solution from the stored bag into 225 mL of BHI broth, incubate for 24 h at 37 °C, streak in a selective agar plate for the chosen BI, and incubate for 18–24 h at 37 °C [4]. For *Enterococcus faecium* NRRL B-2354, the suggestion of selective agar medium is the m-*Enterococcus* agar. For samples subjected to enrichment, the resulting plates are not considered for enumeration. Instead, they should be assessed qualitatively for the presence of typical colonies (*see* **Note 5**).

### 3.4 Data Analysis and Documenting

#### 3.4.1 Determination of Decimal Reduction Number

1. *Number of survivors (N):* Calculate the log of the number of survivors (N) using Eq. 1. Calculate for each time point in all the replicated processes:

$$\log N = \log \frac{\sum C \cdot D}{V} \tag{1}$$

where $C$ is the average number of colonies at a given time point; $D$ is the inverse of decimal dilution of the chosen plate for count; and V is the volume of the decimal dilution added in the the plate. The concentration of cells at the initial time (before processing) is denominated $\log N_0$, and the concentration at the last time point (after processing) is $\log N_f$.

2. *Real decimal reduction number* ($\gamma_{real}$): Calculate $\gamma_{real}$ using Eq. 2:

$$\gamma_{real} = \log N_0 - \log N_f \tag{2}$$

3. After obtaining $\gamma_{real}$ of all replicates, calculate the mean, the standard deviation (SD), and the confidence interval (CI) applying ANOVA test for a confidence level of 95%.
4. *Minimum Reduction Case (MRC):* Calculate MRC to evaluate the lowest reduction in a worst-case scenario. Select the lowest number of $\log N_0$ and the highest number of $\log N_f$, regardless of which replicate each value comes from. Calculate MRC using Eq. 3 [1]:

$$MRC = \min(\log N_0) - \max(\log N_f) \tag{3}$$

### 3.5 Documenting and Revalidation

1. Document all the obtained information, including the procedure, colony counts, and calculations.
2. The data will be used to check the time–temperature of the designed process, as described in Chap. 17 (*Designing and Checking Static Thermal Processes of Foods*).
3. A revalidation study must be conducted under the following circumstances: (1) Changes in the formulation of the food product; (2) Process failures for which the root cause could not be identified; (3) Changes in hazards associated with the food or its ingredients (e.g., the identification of a new hazard or the detection of a higher concentration of microbial cells than that was originally considered during the process design) [1].

## 4 Notes

1. For example, if the thermal process aims to cause a 5-log reduction, the concentration of cells in the food matrix needs to be $10^7$ CFU/g or mL (7 log CFU/g or mL).
2. The recommended options are mesh bags and wire baskets [2]. See an example of a container for almonds in Fig. 2
3. For processes that demonstrate very much variability in cause reduction, it is important to conduct five or more replicates [1].
4. To know how to use thermocouples, refer Chap. 13 (*Temperature Measurements Using Thermocouples*).

**Fig. 2** Inoculated almond test cage [5]

5. For cases that require enrichment, the results will be expressed as qualitative values (presence/absence) [6]. If the detection of survivors in samples requires only this type of method, the quantity of samples must be larger (10 to 25 samples) [1].

## References


1. Ceylan E, Amezquita A, Anderson N, Betts R, Blayo L, Garces-Vega F *et al* (2021) Guidance on validation of lethal control measures for foodborne pathogens in foods. Compr Rev Food Sci Food Saf 20(3):2825–2881
2. Almond Board of California. Guidelines for Using *Enterococcus faecium* NRRL B-2354 as a Surrogate Microorganism in Almond Process Validation [Internet]. 2014. Available from: https://www.almonds.com/sites/default/files/2020-05/guidelines_for_using_enterococcus_faecium_nrrl_b-2354_as_a_surrogate_microorganism_in_almond_process_validation.pdf
3. Campden BRI (2018) The application of validation principles to continuous thermal and non-thermal processing. Available from: https://www.campdenbri.co.uk/white-papers/continuous-processing.php
4. Channaiah LH, Holmgren ES, Michael M, Sevart NJ, Milke D, Schwan CL *et al* (2016) Validation of baking to control *Salmonella* Serovars in hamburger bun manufacturing, and evaluation of *Enterococcus faecium* ATCC 8459 and *Saccharomyces cerevisiae* as nonpathogenic surrogate indicators. J Food Prot 79(4): 544–552
5. Almond Board of California. Rotary Roaster Validation Guidelines for Industry. Available from: https://www.almonds.com/sites/default/files/2023-03/Rotary%20Roaster%20Validation%20Guidelines.pdf
6. Diep B, Moulin J, Bastic-Schmid V, Putallaz T, Gimonet J, Valles AD *et al* (2019) Validation protocol for commercial sterility testing methods. Food Control 103:1–8

# Chapter 17

# Designing and Checking Static Thermal Processes of Foods

**Marcelo Felipe Silva Estácio Santana**

## Abstract

Thermal processes are used to control enzymes and food spoilage and pathogenic microorganisms, assuring food safety and shelf-life stability. Designing these processes requires an understanding of the equipment operation, data on the heat penetration of food, and the behavior of target microorganisms during the thermal process expressed in the thermal inactivation parameters. The aim of designing thermal processing is to determine the time–temperature of a process lethal to a specific microorganism, achieving safe levels and resulting in diminished changes in physical–chemical properties. Commonly employed methods for such calculations are the general method, based on Bigelow's equation, and mathematical methods, such as Ball's formula. This chapter shows the steps to determine $F_p$ using both methods and the final steps to check the thermal process.

**Key words** Thermal validation, Bigelow's equation, Ball's formula

## 1 Introduction

From the point of view of microbiology and food safety, a thermal process is a treatment based on the application of heat that aims to inactivate enzymes and reduce the population of spoilage and/or pathogenic microorganisms [1]. Although this looks like a simple definition, designing a thermal process and ensuring its safety demand some effort. A thorough understanding of the equipment and its operation, along with the knowledge of microbial hazards, food composition, and microbial behavior under heat treatment, is essential for designing an effective thermal process. This ensures the selection of appropriate treatment, avoiding both sub- and overprocessing.

For knowledge purposes, sub-processed food is dangerous once the time–temperature relation is not enough to guarantee the reduction of the population of cells to a safe level, which can elevate the risks of illness to the consumer or spoilage of the food. On the other hand, overprocessing impacts negatively on nutrients

Anderson S. Sant'Ana (ed.), *Heat Resistant Microorganisms in Food Safety and Spoilage*, Methods and Protocols in Food Science, https://doi.org/10.1007/978-1-0716-5280-0_17,

(such as vitamins and bioactive compounds), color, texture, and flavor of the food, leading to rejection of the product by consumers.

Therefore, this chapter demonstrates the steps to design and check static thermal processes (food heat treatment conducted in batches). The study focused on demonstrating two methods commonly used in designing thermal processes: the general method and Ball's formula method. Beyond that, the objective of this work was to evaluate the impacts of thermal process design on the reduction of microbial population, not including a description of methods and calculations about the degradation kinetics of nutrients and physical changes.

## 2 Materials

Most of the data required to design the static thermal process is obtained from other steps described in previous chapters of this book:

1. Chapters 9 and 10 describe the materials and methods to obtain microbiological data (D and z values and other thermal inactivation parameters).
2. Chapter 13 describes how to obtain the temperature in the interior of the food in a static thermal process.
3. Chapters 15 and 16 describe the preparation of biological indicators, their inoculation into food products, and the procedures for conducting a validation study.

Beyond the data obtained through the protocols observed in previous chapters, the design and check of the thermal process requires the utilization of software to conduct the calculation steps. Suggestions are Microsoft Office Excel, PTC Mathcad, and others.

## 3 Methods

Figure 1 is a simple demonstration of the steps to design and check a thermal process.

### 3.1 Selecting Variables and Objectives

1. *Selecting the hazard*: A thermal process is designed to achieve reductions of a specific microorganism or enzyme to safe levels for the consumer or the food product. In this way, it is necessary to identify the actual threat by investigating the prevalence of the hazard in the specific food, its ingredients, and similar products, reviewing historical outbreaks associated with it, and consulting current legislation. Generally, the thermal process for low-acid food (pH >4.5) has as its target pathogens, such as

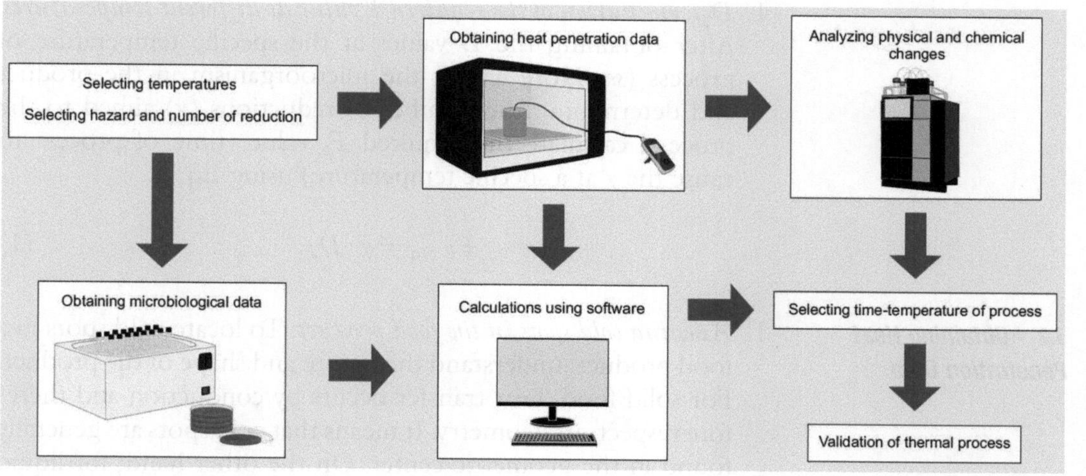


**Fig. 1** Scheme of the steps to design and check a static thermal process. (The author)

*Clostridium botulinum* spores. In acid food products (pH <4.5), the hazards are associated with food spoilage, e.g., spores of *Geobacillus stearothermophilus*, *Alicyclobacillus acidoterrestris*, yeasts, molds, etc. [1]. Therefore, conclusions regarding hazards targeted by thermal processing also require knowledge of the physicochemical properties of the product.

2. *Determining the number of reductions*: All thermal processing has a defined number of reductions ($\gamma$) of the target microorganism population that must be achieved to guarantee safety. For example, in commercial sterility, where *Clostridium botulinum* is the target, $\gamma$ expected is 12 log-cycles ($\gamma = 12$) [2]. For acid and high-acid food, $\gamma$ is not well established. In this case, search for legislation or regulations that specify a $\gamma$ or find databases about microorganisms in foods and ingredients, such as typical concentrations, minimum population responsible for causing illness or spoilage, and other relevant information that helps determine a $\gamma$.
3. *Microbiological data:* After selecting the target and the $\gamma$, determine its thermal inactivation parameters in the food matrix at different temperatures. Search the literature for D- and z-values in the same food matrix or products with similar composition, considering factors such as water activity ($a_W$), fat content, and other relevant parameters. If there is no information about these parameters in the food matrix or similar products, it is important to determinate them. In this case, refer Chaps. 10 and 11 (*Estimating Inactivation Kinetic Parameters of Microorganisms: Log-Linear Kinetics* and *Estimating Inactivation Kinetic Parameters of Microorganisms: Non-Log-Linear Kinetics*, respectively) to understand how to estimate the inactivation kinetic parameters of microorganisms.

4. *Determination of the required F value at different temperatures:* After obtaining the $D$ values at the specific temperature of process (*see* **Note 1**) for the microorganism in the product and determining the number of reductions ($\gamma$) aimed to the process, calculate the required $F_p$ value (time of process to cause the $\gamma$ at a specific temperature) using Eq. 1.

$$F_{P\ \mathrm{req}} = \gamma \cdot D_T \tag{1}$$

### 3.2 Obtaining Heat Penetration Data

1. *Assessing cold spots in the food product*: To locate cold spots in a food product, understand the nature and shape of the product. For solid food, heat transfer occurs by conduction and therefore respects its geometry. It means that cold spots are generally found in the geometric center. On the other hand, for liquid placed in a container (can, bottle, etc.), heating promotes agitation, resulting in convection heat transfer, which makes cold spots located in different regions of the container, especially in static points where there is no convection [3]. So, the heat transfer is through conduction, which results in a slow temperature rise.
2. *Measuring temperature profile in cold spots of the food product*: After determining the time and temperature of the process and finding the cold spots, conduct the thermal process for a food product. Before introducing the food into the heater, insert thermocouples in the cold spots to measure the increase and decrease of the temperature of the product during the thermal process to obtain a temperature profile. An example of a temperature profile during a thermal process is shown in Table 1 and Fig. 2. To understand how to measure the temperature in the cold spots of a food product, refer Chap. 13 (*Performing Heat Penetration Studies*).

### 3.3 Calculations

#### 3.3.1 General Method

1. *Determination of the lethality rate (L):* After obtaining the temperature profile, calculate the lethality rate (L) at each time point (i), using Bigelow's formula (Eq. 2) [4]:

$$L_i = 10^{\frac{(T_i - T_{\mathrm{ref}})}{z}} \tag{2}$$

where $T_i$ is the temperature at a given time point $i$; and $T_{\mathrm{ref}}$ is the reference temperature of the process in °F (*see* **Note 2**).

To better observe the behavior of the lethality rate over the time of processing, graphics are indispensable. Therefore, Fig. 3 shows a graphic made in Microsoft Office Excel as an example.

2. *Calculating $F_p$:* Integrating Bigelow's formula over thermal processing time (Eq. 3) enables determination of the accumulated lethality ($F_p$), also known as sterilization process equivalent time:

**Table 1**
**Fictitious example of internal temperature profile of a food product during a treatment at 250 °F for 15 min**

| Stage | Time (min) | Autoclave temperature (°F) | Food temperature (°F) |
|---|---|---|---|
| Come-up time | 0 | 70.5 | 77 |
| | 2 | 170.6 | 79 |
| | 5 | 250.1 | 84.2 |
| Thermal process | 5 | 250.1 | 84.2 |
| | 10 | 250.3 | 132.8 |
| | 12 | 250.9 | 176 |
| | 15 | 251.6 | 208.4 |
| | 20 | 251.6 | 244.4 |
| Cooling | 20 | 251.6 | 244.4 |
| | 25 | 228.2 | 234.5 |
| | 30 | 176 | 212 |
| | 35 | 105.8 | 176 |
| | 40 | 71.6 | 104 |

The author

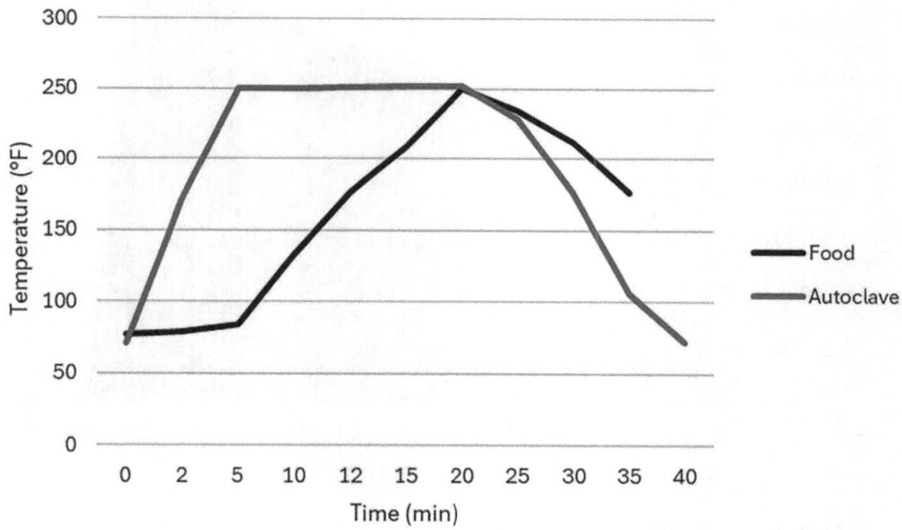


**Fig. 2** Temperature vs time at food cold spot and autoclave during thermal process (250 °F/15 min). (The author)

$$F_p = \int 10^{\frac{(T - T_{\mathrm{ref}})}{z}} \mathrm{d}t \qquad (3)$$

Since the behavior of the lethality curve typically requires numerical methods to calculate $F_p$ [5], one commonly used approach is the trapezoidal rule. In this method, the area under

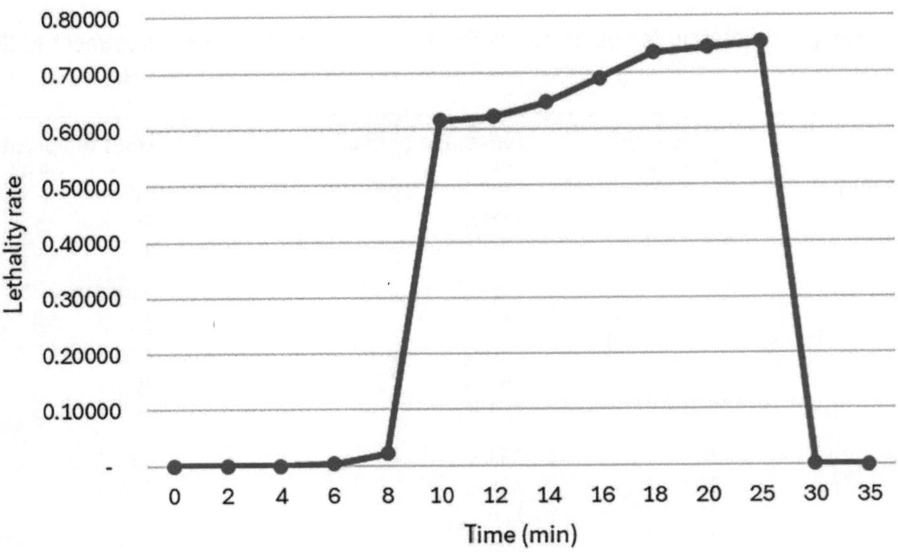


**Fig. 3** Lethality rate vs time of processing. (The author)

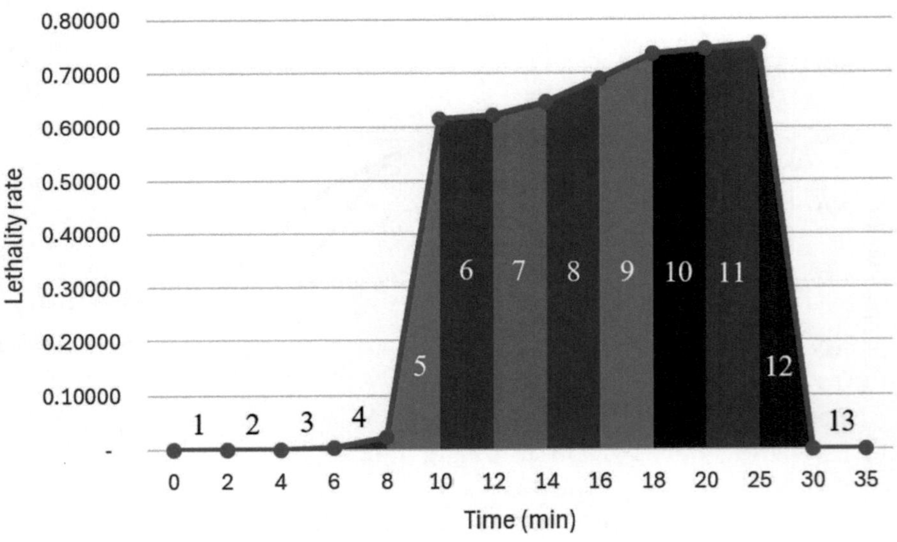


**Fig. 4** Areas of lethality curve. (The author)

the lethality curve is divided into smaller trapezoidal segments, which can then be calculated using Eq. 4. Figure 4 shows an example using the curve found in Fig. 3:

$$\Delta F_{p_i} = \frac{L_i + L_{i-1}}{2} \cdot \Delta t \tag{4}$$

where $\Delta F_{p_i}$ is the lethality of the interval between the given time-point $i$; $\Delta t$ is the time interval between the given time-point $i$.

3. *Calculating $F_p$:* Next, calculate the sum of all $\Delta F_{p_i}$ values to obtain the total $F_p$ (Eq. 5):

$$F_p = \sum_{i=1}^{n} \Delta F_{p_i} = \Delta F_{p1} + \Delta F_{p2} + \ldots + \Delta F_{p_n} \quad (5)$$

### 3.3.2 Ball's Formula Method

1. Ball's formula (Eq. 6) [6] is also widely used for the determination of $F_p$. For that, data obtained from the study of heat penetration (*see* Chaps. 13 and 14) are also needed:

$$\log g = \log(T_a - T_g) = -\frac{t_b}{f_h} + \log[j_h(T_a - T_0)] \quad (6)$$

where $T_a$ is the temperature of the heating medium; $T_g$ is the temperature of food at the end of the thermal process; $t_b$ is the time of Ball, expressed in Eq. 7; $f_h$ is the heating rate; $j_h$ is the lag factor; and $T_0$ is the initial temperature of the product.

$$t_b = t_r + 0.42 \cdot t_{\mathrm{CUT}} \quad (7)$$

2. *Determining $f_h$ and $j_h$:* Chap. 14 (*Estimating Heat Penetration Parameters*) describes how to obtain the parameters of lethality, $f_h$ and $j_h$, using the heat penetration data and the curves of heating and cooling.
3. *Determining log g and U:* With all necessary parameters and data available (*fh*, *jh*, *tb*, *Ta*, and *To*), log g can be calculated using Eq. 6. Then, using the tables provided by Vinters et al. [7], which correlate log g with R (the ratio between fh and U), it is possible to determine the sterilization value of the thermal process using Eq. 8:

$$R = \frac{f_h}{U} \quad (8)$$

4. *Determining $F_p$:* Finally, find $F_p$ using Eq. 9:

$$F_p = U \cdot L = U \cdot 10^{(T_a - T_{\mathrm{ref}})} \quad (9)$$

## 3.4 Selecting and Checking the Process

1. *Comparing the results:* Compare all results of $F_p$ obtained using two methods at the same temperature. To guarantee the aimed reduction and avoid errors, $F_p$ obtained by the methods must be higher than $F_{p\ \mathrm{req}}$ calculated based on the aimed $\gamma$ (*see* **Note 3**).
2. *Selecting time–temperature of the process*: It is observed that the higher the temperature of the process, the shorter the time required to inactivate the microorganisms in the product. However, applying harsh thermal treatment implies

undesirable chemical and physical reactions, resulting in negative changes in nutrient content, texture, color, and flavor of the food, which impact its commercialization. On the other hand, employing mild temperatures to avoid such problems implies longer thermal processing time, which is not economically advantageous to the producer. In the face of this, it is indispensable to determine the kinetics of food quality changes and perform laboratory analysis to evaluate the physicochemical changes, to select the optimal time–temperature parameters for the thermal process [5].

3. *Validation*: After the design, the thermal process must be checked. Therefore, it is important to conduct a study of validation, using the target microorganism, surrogate or a biological indicator. See Chap. 16 (*Conducting a Validation Study of Thermal Processing with Biological Indicators*) to know how to conduct such a study.

## 4 Note

1. A precise margin percentage between $F_p$ calculated by methods and the required $F_{p\ req}$ is not well established. However, in practice, values of $F_p$ that are at least 20% higher than $F_{p\ req}$ are frequently adopted.

### References

1. Ramesh MN (2020) Pasteurization and food preservation. In: Rahman MS (ed) Handbook of food preservation, 3rd edn. CRC Press, Boca Raton, pp 599–608
2. Stumbo CR, Purohit KS, Ramakrishnan TV (1975) Thermal process lethality guide for low-acid foods in metal containers. J Food Sci 40(6):1316–1323
3. Holdsworth D, Simpson R (2007) Heat transfer. In: Thermal processing of packaged foods. Springer, Boston, pp 14–86
4. Bigelow WD (1921) The logarithmic nature of thermal death time curves. J Infect Dis 29(5): 528–536
5. Stoforos NG (1995) Thermal process design. Food Control 6(2):81–94
6. Ball CO (1923) Thermal process time for canned food. Bulletin of the National Research Council, p 46
7. Vinters JE, Patel RH, Halaby GA (1975) Thermal process evaluation by programmable computer calculator. Food Technol J 29(3):44–48

# Chapter 18

# Designing and Qualifying Continuous Thermal Processes of Foods

**Izael Gressoni Junior and Alessandra Regina da Silva**

## Abstract

Continuous thermal processing is one of the most reliable technologies for ensuring microbial safety and extended shelf life of fluid foods while preserving their nutritional and sensory attributes. This chapter provides a systematic framework for designing and qualifying continuous flow heat treatments by integrating microbial inactivation kinetics with fluid dynamics. It highlights the essential steps of food matrix characterization, selection of microbial targets, and determination of critical kinetic parameters such as D- and z-values and F-value, combined with hydrodynamic analysis based on Reynolds number for Newtonian and non-Newtonian fluids. Emphasis is placed on the accurate prediction of minimum residence time ($F_{min}$) to guarantee the safety margin expressed as the Probability of a Non-Sterile Unit (PNSU). The chapter also discusses experimental and computational tools and modeling strategies to achieve both compliance and product quality in industrial operations.

**Key words** Continuous thermal processing, Microbial inactivation kinetics, Residence time distribution (RTD), Newtonian and non-Newtonian fluids, Reynolds number, Decimal reduction time (D-value), z-value, F-value

## 1 Introduction: Continuous Thermal Process Calculations in Foods

Food industries hold the responsibility for designing and implementing scientifically robust food safety management systems, capable of controlling biological, chemical, and physical hazards in compliance with current regulatory frameworks. Thermal processing, nonthermal technologies, or an integrated hurdle approach should be applied to ensure the reduction of foodborne hazards to levels that do not pose public health risks under standard conditions of distribution and storage [1]. Various processing technologies are available to render foods microbiologically stable and safe for extended shelf life. The appropriateness of each system is determined by the specific characteristics of the food matrix, production throughput, capital investment capacity, and labor costs. In system selection, product quality attributes and economic feasibility

Anderson S. Sant'Ana (ed.), *Heat Resistant Microorganisms in Food Safety and Spoilage*, Methods and Protocols in Food Science, https://doi.org/10.1007/978-1-0716-5280-0_18,

represent the primary decision-making factors. Given that overhead and labor constitute the major operational expenses, facilities with high processing capacity tend to adopt systems with elevated capital requirements but reduced labor dependency. Superior product quality is generally achieved with technologies capable of delivering high-temperature, short-time (HTST) treatments, which optimize microbial inactivation while preserving sensory and nutritional properties. A precise comprehension of the mechanisms governing microbial thermal inactivation is indispensable for the establishment and validation of safe heat-processing protocols in the food industry. The lethality of thermal treatments is described by kinetic parameters of microbial death, particularly the decimal reduction time (D-value), which quantifies the time required at a constant temperature to reduce the microbial population by one logarithmic cycle, and the thermal resistance constant (z-value), which indicates the temperature increase necessary to achieve a tenfold reduction in the D-value. These parameters are central to predictive modeling of microbial inactivation. Thermal inactivation of microorganisms is recognized as a fundamental unit operation in food preservation, as it not only ensures product stability and extended shelf life but also guarantees compliance with microbiological safety standards. Understanding these kinetic aspects allows engineers to optimize time–temperature combinations, balancing microbial lethality with the preservation of nutritional and sensory quality attributes [2]. In continuous thermal processing applied to the food industry, it is mandatory that the product be maintained at a defined target temperature for a predetermined residence time. The control of this *time–temperature relationship* is decisive, as it dictates the magnitude of microbial lethality and enzymatic inactivation, while also influencing physicochemical and sensory transformations of the food matrix. From a kinetic perspective, microbial inactivation under continuous flow systems is modeled using *D-values* and *z-values*, which allow the prediction of log-reductions achieved under specific conditions. In parallel, enzymatic denaturation exhibits different activation energies and thermal sensitivities, requiring careful adjustment of process parameters. Thus, the design of continuous systems—such as *plate heat exchangers, tubular sterilizers, or scraped-surface heat exchangers*—must balance the achievement of the desired microbial log-reduction with the preservation of nutritional compounds and sensory attributes. The challenge lies in optimizing flow dynamics and heat transfer to ensure both *process lethality* and *product quality* in large-scale operations. The prediction of holding time in continuous thermal systems is complex due to the nonuniform residence time distribution (RTD) of fluid elements along the tube. This variability arises from the differences in velocity profiles within the conduit, meaning that not all fluid particles are exposed to the same time–temperature conditions.

Classical experimental investigations [3] on velocity profiles in pipe flow and turbulence, establishing the relationship between the Reynolds number and velocity distribution, demonstrated that the velocity distribution across a circular tube is directly dependent on the *Reynolds number (Re)* under conditions of fully developed flow. The classical findings established that the velocity ratio—defined as the quotient of bulk-averaged velocity to the maximum velocity at the tube center—varies linearly with the logarithm of Reynolds number in turbulent regimes, regardless of tube diameter. The *Reynolds number* is a key dimensionless parameter used to classify flow regimes in fluid dynamics. In laminar flow, typically observed at Reynolds numbers below 2,100, the velocity distribution is highly ordered and predictable. As the Reynolds number increases, a transitional range occurs between 2,100 and 10,000, where the flow alternates between laminar and turbulent instabilities. A fully turbulent flow is established above 10,000, characterized by intense mixing and improved heat transfer. For non-Newtonian fluids, such as starch dispersions, emulsions, or protein-based systems, the classical Reynolds number must be replaced by a generalized form that accounts for rheological properties like flow behavior index and consistency coefficient. This adjustment is essential for accurately predicting flow regime, velocity distribution, and residence time distribution. Such parameters are critical for the design and validation of thermal processes in the food industry, where both microbial safety and product quality depend on the precise control of fluid dynamics. From a food engineering perspective, these hydrodynamic principles are critical because the *residence time distribution (RTD)* directly influences the extent of microbial inactivation and enzymatic denaturation. In turbulent flow, zones of faster or slower moving fluid lead to heterogeneity in thermal exposure, potentially creating *cold spots* that reduce process lethality. Therefore, accurate modeling of RTD and velocity profiles is fundamental in the design and validation of *continuous thermal systems* such as HTST pasteurizers and UHT sterilizers, ensuring that all fluid elements receive adequate thermal treatment for food safety.

In light of the critical importance of accurately establishing thermal processes within the food industry, this chapter seeks to provide a comprehensive introduction to the general principles of thermal process design, with particular emphasis on the development and validation of continuous thermal processing systems.

## 2 Materials

The establishment and validation of continuous thermal processes in foods require a diverse set of materials that can be grouped into different categories:

### 2.1 Food Matrices and Model Systems

- Raw materials and finished food products with *characterized physicochemical and rheological properties* such as density, viscosity, thermal conductivity, and specific heat.
- *Model fluids* (Newtonian and non-Newtonian) are frequently employed to simulate industrial formulations and to evaluate flow behavior and residence time distribution (RTD) under controlled conditions.

### 2.2 Processing Equipment

- *Pilot- and laboratory-scale heat exchangers*, including tubular, plate, and scraped-surface systems, to reproduce heating, holding, and cooling operations under conditions analogous to industrial processes.
- *Holding tubes* with well-defined geometry and length, which are essential for evaluating time–temperature exposure of fluid elements.

### 2.3 Measuring Instruments

- *Calibrated thermocouples and resistance temperature detectors (RTDs)* for accurate temperature monitoring at multiple points of the system.
- *Flow meters and pressure sensors* to quantify volumetric flow rates, velocity distribution, and pressure drop along the equipment.
- *Data loggers and acquisition systems*, which allow continuous monitoring and recording of thermal and hydrodynamic parameters during processing.

### 2.4 Microbiological and Biochemical Indicators

- *Target microorganisms*, typically *salmonella* spp. or validated surrogates such as *enterococcus faecium* NRRL B-2354, to assess microbial lethality under defined conditions.
- *Enzymatic markers*, such as peroxidase or alkaline phosphatase, to evaluate the extent of enzyme inactivation as complementary quality indicators.

### 2.5 Analytical and Computational Tools

- *Laboratory assays* for microbial enumeration, enzyme activity, and physicochemical property determination.
- *Computational models and software* for predicting thermal inactivation kinetics, residence time distribution, and process lethality, based on parameters like D- and z-values and flow characteristics (including generalized Reynolds number for non-Newtonian fluids).

Together, these materials provide the experimental and analytical basis for the design, optimization, and verification of continuous thermal processes in the food industry, ensuring that both *food safety objectives* and *quality preservation criteria* are achieved.

## 3 Methods

### 3.1 General Calculations on Food Thermal Processes

#### 3.1.1 Physicochemical Characterization of the Food Matrix

To establish safe thermal processes, the primary consideration is the product specification, as thermal treatments are inherently product-dependent. Each food matrix therefore requires its own specific and nontransferable thermal process. Key physicochemical parameters to be determined include pH, water activity (aw), soluble solid content (°Brix), density, and viscosity—variables that are essential, for instance, in the calculation of the Reynolds number (*see* Subheading 3.2.2).

Furthermore, it is critical to quantify the initial population of the target microorganism in the newly formulated product, considering the batch or lot in its entirety.

#### 3.1.2 Determination of the Microbiological Target

Considering that thermal processes are unique to each food, the microbiological target must be determined from the ingredients that make up the product or from the formulated product, immediately before the thermal process. In general, the heat-resistant spoilage spores (aerobic and anaerobic) present in the different raw materials or in the formulated product are enumerated, considering these as the most heat-resistant form (the most difficult to eliminate). Thus, once these are eliminated, the pathogenic targets, which are generally less resistant, will also be eliminated. In this sense, then, for thermal process establishments, it is always better to use the product-specific method.

#### 3.1.3 Establishment of Thermal Inactivation Kinetics

The determination of the maximum thermal resistance of potential spoilage microorganisms, as well as the resistance of the specific microorganisms naturally contaminating the product, is essential for establishing safe processing parameters. Kinetic factors must be quantified under controlled laboratory conditions using the actual food matrix as the experimental medium. For Pflug [4], comparative analyses between empirical data accumulated over more than a century of food processing and laboratory-derived D-values for mesophilic or thermophilic spore-forming organisms indicate that the maximum D-values obtained under laboratory conditions can be two to three times higher than the effective D-values observed in naturally contaminated.

products.

According to Stumbo [5], the most common methods used for measuring thermal resistance of bacteria may be roughly classified as follows:

- Thermal Death Time (TDT)—tube method [6]
- Thermal Death Time (TDT)—can method
- “Thank” method
- Flask method

**Table 1**
**Moist-heat resistance characteristics of important spoilage microorganisms [4]**

| Microorganism | D-value (min) |
|---|---|
| *Bacterial spores* | |
| *Geobacillus stearothermophilus* | $D_{121.1°C} = 3.5$ |
| *Bacillus coagulans* | $D_{121.1°C} = 1.5$ |
| *Clostridium sporogenes* | $D_{121.1°C} = 1.0$ |
| *Bacillus subtilis* | $D_{121.1°C} = 0.3$ |
| *Clostridium botulinum* | $D_{121.1°C} = 0.2$ |
| *Fungal spores* | |
| *Aspergillus niger* (conidia) | $D_{60.0°C} = 1.0$ |
| *Byssochlamys fulva* (ascospores) | $D_{90.0°C} = 1.0$ |
| *Vegetative cells* | |
| Food spoilage organisms | $D_{80.0°C} = 1.0$ |
| Viruses | $D_{75.0°C} = 1.0$ |
| Pathogenic organisms | $D_{65.0°C} = 1.0$ |

Source: Pflug [4]

- Thermoresistometer method
- Unsealed TDT tube method
- Capillary tube method

All of these methods, and some modifications (i.e., use of stand-up pouches), are being used at present to obtain data for thermal process calculation. The selection of the most appropriate method will depend on the rheological characteristics of the product: For very viscous products, stand-up pouches should preferably be used, and for more liquid products, TDT or capillary tubes, for example. In addition to selecting the appropriate method for thermal destruction studies, another very important step is the analysis of survival data. In this sense, there are methods that consider linear death (40% of thermal inactivation curves) and methods that consider non-log-linear thermal inactivation (60% of inactivation curves), with the formation of shoulders or tails at the beginning or end of heating, respectively. The main methods for analyzing thermal inactivation data would be:

- *Linear models*

  (a) Bigelow and Esty [6]: This model assumes that all the cells in a population have equal heat sensitivity.

$$\log_{10}(N) = \log_{10}(N_0) - \frac{t}{D} = \log_{10}(N_0) - \frac{k_{\max}t}{\ln(10)} \quad (1)$$

where

*N*: Final population, spore/mL

*N*(0): Initial population, spore/mL

$K_{\max}$: The first-order inactivation constant, 1/time unit

*D*: Decimal reduction time, time unit

- *Log linear models with shoulder or tailing*

The proposed model is capable of describing microbial inactivation through a log-linear pattern that may include or exclude the presence of shoulder and/or tailing regions, thereby enabling a continuous and smooth transition between these phases. Within this modeling framework, shoulder is defined as an initial population more resistant to heat effects, and tailing is defined as a residual subpopulation that remains stable over time, exhibiting no appreciable further inactivation. This interpretation fundamentally differs from that of the biphasic model, in which the apparent tailing reflects the sequential inactivation of two distinct microbial subpopulations with different resistance levels:

$$\log_{10}(N) = \log_{10}\left[\left(10^{\log_{10}(N_0)} - 10^{\log_{10}(N_{res})}\right).e^{-k_{\max}t}.\left(\frac{e^{k_{max}S_1}}{1+(e^{k_{\max}S_1}-1).e^{-k_{\max}t}}\right) + 10^{\log_{10}(N_{res})}\right] \quad (2)$$

With log 10(*N*(0)), log 10($N_{res}$), $k_{\max}$, and Sl being the degrees of freedom used for the parameter estimation by GInaFiT [7].

- *Weibull-type models*

When employed to describe microbial inactivation kinetics [8], the Weibull model represents the cumulative form of an asymmetric Weibull probability density function characterizing the distribution of individual cell heat resistances. Alternative statistical formulations, such as the symmetric Fermi distribution, can also be applied [9]. This class of models departs fundamentally from the mechanistic. In particular, the presence of a shoulder is not postulated a priori as a discrete physical phenomenon but rather emerges intrinsically from the statistical properties of the resistance distribution—specifically, when the mean of the distribution is large relative to its variance [9]:.

$$\log_{10}(N) = \log_{10}(N_0) - \left(\frac{t}{\delta}\right)^p \quad (3)$$

where *d* [time unit] is a scale parameter and can be denoted as the time for the first decimal reduction if $p = 1$, and p [−] is a shape parameter. For $p > 1$, convex curves are obtained, while for $p < 1$, concave curves are described.

- *Biphasic models*

Cerf [10] proposed a two-fraction model, which can be formulated as follows:

$$\log_{10}(N) = \log_{10}(N_0) + \log_{10}\left(f.e^{-k_{\max 1}t} + (1-f).e^{-k_{\max 2}t}\right) \quad (4)$$

where f is the fraction of the initial population in a major subpopulation, (1_f) is the fraction of the initial population in a minor subpopulation (which is more heat-resistant than the previous one), and $k_{\max1}$ and $k_{\max2}$ [1/time unit] are the specific inactivation rates of the two populations, respectively.

Table 1 shows standard moist-heat resistance characteristics, D value, of important spoilage microorganisms. *It is possible to observe that vegetative microorganisms are normally inactivated at temperatures below 90 °C and spore-forming, normally requiring temperatures above 100 °C.*

#### 3.1.4 Establishing the Initial Population—No

Considering the population with the highest heat survival rate recovered from the ingredients or formulated product, it is necessary to convert spores per mL of product to spores per package or batch (shown in calculation example 1). For fluid foods, the initial population will be the product of the number of microorganisms or spores/volume of incoming fluid multiplied by the volumetric flow rate—Eq. 5 (shown in calculation example 2):

$$N_0 = \pi R^2 n_0 \overline{V} \quad (5)$$

1. where

$N_o$ = Number of spores or microorganisms entering the tube

$n_o$ = Number of microorganisms or spores/volume of incoming fluid

$R$ = Tube radius

$\overline{V} = V_{avg}$ = Average velocity

#### 3.1.5 Endpoint of the Microbial-Control Process—N and PNSU

A primary challenge in the design of sterilization processes lies in the precise definition of the *process endpoint*, which must be established prior to the calculation of the *F-value, the cumulative thermal lethality required to achieve the specified level of microbial safety* [4].

The process endpoint represents the *intended level of consumer protection* and is expressed, for an individual product unit, as the *Probability of a Non-Sterile Unit (PNSU)*. For large production lots—on the order of millions or billions of units—this endpoint can also be described as the maximum number of nonsterile units that is statistically acceptable within the lot [4, 5].

According to Pflug [4], in the food industry, PNSU targets are commonly set at probabilities such as *$10^{-4}$, $10^{-6}$, or $10^{-9}$*, which define the required stringency of the sterilization process. For example, a PNSU of $10^{-6}$ theoretically implies that, on a statistical basis, ***no more than one nonsterile container would be expected per one million processed containers.*** Because such low probability levels cannot be directly measured, the development and validation of microbial control processes rely on ***mathematical modeling*** or *experimental approaches* to estimate the process duration and conditions necessary to achieve PNSU levels of $10^{-6}$ or $10^{-9}$.

### 3.1.6 F-Value

When subjected to heat, bacteria die in an orderly fashion, and the order of death is believed to be logarithmic, represented by the general equation Eq. 6:

$$t = D(\log a - \log b) \tag{6}$$

in which

$t$ = Time, in minutes, of heating at a constant lethal temperature.

$D$ = Time, in minutes, to kill 90% of the cells/spores, in a population.

$a$ = Initial number of cells/spores in the population.

$b$ = Number of cells/spores in the population after time $t$.

The term in parentheses in Eq. 6 can be rewritten as follows:

$$\gamma = \log(a) - \log(b) \tag{7}$$

where $\gamma$ represents the number of decimal reductions.

So, rewritten Eq. 6 will be

$$t = D.\gamma \tag{8}$$

The term "$t$" can be considered as the total process lethality, at constant temperature, and was designated by the symbol "F" [5]. Because of product and process variables, F values will vary slightly. Thus, Eq. 9 expresses the formula for calculating the F value of thermal processes:

$$F = D.\gamma \tag{9}$$

in which:

$F$ = Total time, in minutes, of heating at a constant lethal temperature.

$D$ = Time, in minutes, to kill 90% of the cells/spores, in a population.

$\gamma$ = Number of decimal reductions, obtained from Eq. 7.

At this point, it is important to clarify the difference between the nomenclature F and Fo. The term Fo refers exclusively to the total time of heat application at 121 °C, considering *Clostridium botulinum* spores, with $D_{121°C}$ value = 0.21 min (value determined for the most resistant spores of botulinum), *z* value = 10 °C, and the 12D concept (Stumbo, 1965). Thus, to be safe from a public-health standpoint [4], the slowest-heating zone of low-acid canned food should receive a minimum botulism heat process, usually taken as an Fo of 2.5 min, as shown below:

$$F = D\gamma$$
$$F_o = D_{121°C}(\log 1 - \log 10^{-12})$$
$$F_o = 0.21 \times 12$$
$$F_o = 2.52 \text{ or } 2.5 \quad \text{minutes.}$$

### 3.2 Design of Continuous Flow Thermal Process

#### 3.2.1 Type of Fluid Definition

In the design and validation of continuous thermal processes, accurate characterization of a food fluid's rheological behavior is essential. According to some researchers [11, 12], distinguishing whether a product behaves as a Newtonian or non-Newtonian fluid directly affects the prediction of velocity profiles, residence time distribution, and heat transfer coefficients within tubular systems used for sterilization and pasteurization.

**Newtonian Fluids**

Newtonian fluids are characterized by a constant viscosity, independent of the applied shear rate. In other words, their flow behavior remains linear, and the relationship between shear stress and shear rate follows Newton's law of viscosity. Typical examples in food systems include water, simple sugar solutions, and certain vegetable oils. Because their viscosity remains constant, the prediction of velocity profiles, residence time distribution, and heat transfer in Newtonian fluids is relatively straightforward, which simplifies the design and validation of continuous thermal processes.

**Non-Newtonian Fluids**

Non-Newtonian fluids, in contrast, exhibit viscosity that varies with the shear rate. Their rheological behavior can be categorized into shear-thinning (pseudoplastic), shear-thickening (dilatant), or time-dependent (thixotropic and rheopectic) fluids. Many food products, such as tomato paste, starch dispersions, dairy concentrates, emulsions, and protein-rich formulations, demonstrate non-Newtonian behavior. For these fluids, the velocity distribution inside holding tubes is not parabolic as in Newtonian laminar flow but instead depends on rheological parameters such as the *flow behavior index (n)* and the *consistency coefficient (K)*. This variability complicates the prediction of residence time distribution and microbial lethality in thermal processing.

**Importance in Continuous-Flow Thermal Processes**

In continuous-flow sterilization and pasteurization, precise control of residence time and temperature is essential to guarantee microbial inactivation targets while maintaining product quality. For Newtonian fluids, design calculations are more reliable because flow and heat transfer behavior are well-defined. However, for non-Newtonian fluids, under- or overestimation of residence times can occur if rheological properties are not properly incorporated. This is particularly critical because shear-thinning fluids (the most common in food systems) have flatter velocity profiles, which reduce the gap between maximum and average velocities, directly affecting the estimation of ***minimum residence time*** (***$t_{min}$***) used in thermal process validation.

Consequently, understanding whether a food behaves as a Newtonian or non-Newtonian fluid—and accurately characterize its rheology—is indispensable for the correct design, modeling, and validation of ***continuous thermal processes***, ensuring both microbial safety and the preservation of sensory and nutritional attributes.

### 3.2.2 Reynolds Number (Re) calculation

**Reynolds Number for Newtonian Fluids**

For Newtonian fluids, viscosity is constant and independent of shear rate. The Reynolds number (Re) is a dimensionless parameter used to classify flow regimes, defined as the ratio between inertial and viscous forces:

$$\mathrm{Re} = \frac{\rho\, v\, D}{\mu} \tag{10}$$

where:

$\rho$: fluid density (kg.m$^{-3}$)

$v$: average velocity (m.s$^{-1}$)

$D$: internal diameter of the tube (m)

$\mu$: dynamic viscosity (Pa.s)

For Newtonian fluids in laminar flow (Re < 2100), the velocity profile is parabolic, with the maximum velocity at the tube centerline equal to twice the average velocity. For Re > 10,000, flow is turbulent, with a flatter velocity profile. This classification allows accurate prediction of residence time distribution (RTD) and heat transfer, which are fundamental for designing holding tubes in thermal processes.

**Generalized Reynolds Number for Non-Newtonian Fluids**

Many food products (tomato paste, starch dispersions, protein-rich emulsions) are non-Newtonian and exhibit shear-thinning or shear-thickening behavior. For such fluids, the classical Reynolds equation is inadequate, and the generalized Reynolds number (Reg) must be used:

$$\mathrm{Re}_g = \frac{\rho.(v)^{2-n}.D^n}{8^{n-1}.\left(\frac{3n+1}{4n}\right)^n.k} \tag{11}$$

where:

$n$: flow behavior index (adimensional)

$k$: consistency coefficient (Pa.s$^n$)

$v$, $\rho$ and $D$ have the same meaning as in the Newtonian case

For shear-thinning fluids ($n < 1$), the velocity profile is flatter than Newtonian laminar flow, reducing differences between maximum and average velocity and decreasing the risk of underestimating minimum residence times. For shear-thickening fluids ($n > 1$), the opposite occurs, with sharper velocity profiles increasing variability in RTD.

Accurate estimation of Reynolds number is fundamental for:

1. Predicting flow regime: Determining whether the system operates under laminar, transitional, or turbulent conditions is essential for calculating microbial lethality.
2. Residence Time Distribution (RTD): Newtonian and non-Newtonian behaviors directly affect minimum residence time, which defines the coldest portion of the fluid and therefore the probability of microbial survival.
3. Thermal process validation: Regulatory standards require assurance that all fluid elements receive sufficient thermal exposure. Incorrect modeling of Reynolds number may lead to under-processing (safety risk) or over-processing (quality degradation).
4. Equipment design: The design of holding tubes, pumps, and heat exchangers must account for rheological behavior to achieve consistent heating and aseptic processing.

#### 3.2.3 Residence Time Calculation

The specified process for sterilization in continuously flowing systems is a time of residence in a holding tube, an unheated section of the piping system that leads the fluid from the heat exchangers for heating to the heat exchangers for cooling. A back-pressure valve or a positive displacement timing pump is positioned after the cooler to maintain the pressure within the system at a level needed to keep the product boiling temperature higher than the processing temperature.

After cooling, the sterile product must be handled in a sterile atmosphere; therefore, the process is called *aseptic processing*. The time of residence is set by the volume of the holding tube and the rate of fluid flow delivered by a positive displacement pump.

$$t_{avg} = \frac{A_c L}{Q} \quad (12)$$

where $t_{avg}$ is average fluid residence time, $A_c$ is cross-sectional area of the holding tube, $L$ is length of the holding tube, and $Q$ is volumetric rate of flow.

The average velocity ($V_{avg}. = Q/A_c$) may also be used to calculate the time in the holding tube.

$$t_{avg} = \frac{L}{V_{avg}} \quad (13)$$

In most cases, however, the time of residence of the fastest flowing portion of the fluid is used as the required hold time in the thermal process calculations. This is because the highest probability of survivors from the thermal process is contributed by the section of fluid flowing close to the geometric center of the tube.

The minimum time is:

$$t_{min} = \frac{L}{V_{max}} \quad (14)$$

The maximum velocity for Newtonian fluids in laminar flow is:

$$V_{max} = 2.V_{avg} \quad (15)$$

For power flow fluids in laminar flow:

$$V_{max} = \frac{(3n+1)}{n+1}.V_{avg} \quad (16)$$

For Newtonian fluids in turbulent flow, the following equation was derived [13] for $V_{max}$ as a function of the Reynolds number based on the average velocity:

$$V_{max} = \frac{V_{avg}}{0.00336\ \log(Re) + 0.662} \quad (17)$$

## 3.3 Practical Applications in Food Thermal Processing

This example aims to detail (*step-by-step*) the general base calculation for establishing thermal processes in food.

### 3.3.1 Example 1—General Thermal Process Establishment

(a) *Product characterization*

- Plant-based beverage-added guar gum.
- pH 6.80.
- Process target: Thermophilic aerobic spore-formers.
- Product will be sold at room temperature.

(b) *Establishment of initial population—No*

- Population of thermophilic aerobic spore-formers in the formulated product before the process = $5.65 \times 10^1$spore/ 10 mL.
- Packing capacity: 900 mL.

So, considering the initial population/10 mL and the packing capacity, the initial population/packing will be:

$$5.65 \times 10^1 \text{spore}/10\ \text{mL} \times 900\ \text{mL} = 5.08 \times 10^3 \text{spore/packing}$$

- Production batch size: 234000 L

Considering the spore population/packing and the production batch size, it is possible to obtain the initial population/batch:

5.08 × 103 spore 0.9L
X 234 000L
X = 1.32 × 109 spores/batch → this will be the initial spore population (No) by batch.

(c) *Establishment of Final Population—N*

In the case of tropical preserves (foods packaged in airtight packaging that are stable at room temperature), according to Pflug [4], it is recommended to apply $10^{-6}$ as the probability of nonsterile units (PNSU), that is, the probability of survival of one spore for every 1,000,000 packages produced for thermophiles. This data is used to establish the final population ($N$) that is desired in the process.

Thus:

$N$ = (1 spore/106 packages) × 2.6 × 105 packages/batch → 0.26 thermophilic spore/batch
Or $N$ = 0.26 spore/batch.

(d) *Establishment of Number of Decimal Reductions—γ*

Using the initial and final population values, it is possible to apply Eq. 7 to calculate $\gamma$:

$$\gamma = \log\ (No) - \log\ (N)$$

where:

$\gamma$ = number of decimal reductions.
*No* or $a$ = initial population/batch.
$N$ or $b$ = final population/batch.

$$\gamma = \log\ (1.32 \times 10^9 \text{spore/batch}) - \log\ (0.26\ \text{spore/batch})$$

$\gamma = 9.7$, that is, it is necessary to apply 9.7 decimal reductions to achieve a PNSU of $10^{-6}$.

(e) *Establishment of F-Value*

Once the number of decimal reductions required for the process to stabilize the product has been established, the total process time (F, minutes) to be applied in the retention tube can be calculated, considering a temperature of 135 °C, considering a value of $D_{135°C} = 2.15$ s, using Eq. 8.

$$F = \gamma D$$

where $F$ = total heating time at a constant temperature (min or seconds) for any microorganism, in this case spores of thermophilic aerobic; $\gamma$ = number of decimal reductions required for the process; $D$ = time required for a decimal reduction in the process temperature (min or s).

$$F_{135°C} = 9.7 \times D_{135°C}.$$
$$F_{135°C} = 20.86 \text{ sec.}$$

*Therefore, in this case, the process to be applied to the example product in question would be 21 s at 135 °C, and this time should be applied in the holding tube.*

### 3.3.2 Example 2—Continuous Thermal Process Establishment

This example aims to detail (step-by-step) for the establishment of the continuous thermal process (F minimum, min) for plant-based beverage, considering a 3.0-inch outer diameter retention tube [2, 13–16].

- *Product and process information*
  - System flow rate = 13,500 L/h or 225 L/min.
  - Outer diameter of the holding tube = 3.0 inches.

Following the proposal by Toledo et al. [2], a sanitary tube with an external diameter of 3.0 inches has an internal diameter of 0.07289 m and, therefore, a radius of 0.0364 m.

- Viscosity = 0.04 kg/m.s or 40 mPa.s
- Density = 1.048 g/cm$^3$ or 1048 kg/m$^3$
- Brix = 11°
- pH = 6.4
- Largest population of aerobic sporulating thermophiles (target of thermal process) in the product = $5.65 \times 10^1$ spores/10 mL or 5.65 spores/mL

Based on the information above, **steps 1–5** were applied to establish the minimum retention time, since, according to Toledo et al. [2], the required thermal process time is equal to the minimum residence time in the tube, of the particle that travels the fastest inside it, called "F min," in minutes.

1. *Mean speed calculation ($V_{avg}$; minutes)*

$$V_{\text{avg}} = Q/A$$

where:

Q = flow rate = 225 L/min.

A = cross-sectional area of the tube.

Rearranging:

$$V_{\text{avg}} = \text{flow rate } (L/\min)^* \left(1\,\text{m}^3/1000\ L\right) * (1\ \ \min/60\ \ \text{s}) * \left(1/\left(\pi^*(0.5^*\text{inside diameter})^2.\text{m}^2\right)\right)$$

Substituting the values:

$V_{avg}$ = (225 L/min)*1 $m^3$/1000 L)*(1 min/60 s)*(1/(π*(0.5*0.07289)$^2$.$m^2$).

Considering π = 3.1416 and cancelling units:

$V_{avg}$ = (225)m/(1000*60 s*3.1416*(0.5*0.07289)$^2$).

$V_{avg}$ = 0.904 m/s.

Thus, the mean fluid velocity in the holding tube is 0.0904 m/s.

2. *Calculation of the Reynold's number (Re) and establishment of the fluid regime*

$$\text{Re} = \left(\text{internal diameter}^*\ V_{\text{avg}}{}^*\ \text{density}\right)/(\text{viscosity})$$

where:

- Internal diameter: 0.07289 m.
- $V_{avg}$ = average velocity = 0.904 m/s—calculated in item 1.
- Density = 1.048 g/mL or 1048 kg/$m^3$.
- Viscosity = 0.04 kg/m.s.

Substituting the values, as well as canceling units common to the numerator and denominator:

Re = (0.07289 m * 0.904 m/s * 1048 kg/$m^3$) / (0.04*kg/m.s).

Re = 1726.39.

Therefore, the fluid regime is classified as laminar, since, according to Toledo et al. [2], fluids with Re ≥2100 should be considered as laminar.

3. *Calculation of maximum velocity (Vmax) for laminar flow of Newtonian fluids, according to* Toledo et al. [2].

$$V_{max} = (2 \times V_{avg})$$

where:

- $V_{max}$ = maximum velocity (m/s)
- $V_{avg}$ = average velocity, m/s = 0.904 m/s

Substituting the above values:

$V_{max} = (2 \times 0.904$ m/s)
$V_{max} = 1.81$ m/s

Thus, the maximum speed of the particle traveling fastest through the holding tube is 1.81 m/s.

4. *Calculation of the initial thermophilic aerobic spore population (No) entering the retention tube by cross-sectional area, according to* Toledo et al. [2].

$$No = \pi^* R^2.no.V_{avg}$$

where:

$\pi = 3.1416$

$R^2$ = internal radius = 0.0364 m

no = largest population of aerobic sporulating thermophiles (target of thermal process) in the product = $5.65\times10^1$ spores/10 mL or 5.65 spores/mL

$V_{avg} = 0.0904$ m/s

Thus, substituting the above values:

No = 3.1416*(0.0364 m)$^2$ *(5.65 spores/mL)*0.904 m/s
No = 0.58(spore*m$^3$)/(mL*s) * (10$^6$ mL/m$^3$)
No = 2,03 [(spore*m$^3$) / (s*mL)] * (10$^6$ mL/m$^3$)
No = $5.8\times10^5$ spores/s

Thus, the No entering the holding tube is $5.8\times10^5$ spores/s. This means that the holding tube is entering at $5.8\times10^5$ spores/s, due to the higher initial load of aerobic thermophilic spores found in the plant-based beverage product.

5. *Calculation of the minimum retention time at 135 °C (F min at 135 °C), according to* Toledo et al. [2].

$$\gamma = F_T{}^* D_T \text{ or} \rightarrow F_T = \gamma^* D_T$$

where:

$\gamma$ = number of decimal reductions required = Log (*No*)—Log (*N*)

*No* = initial population (No) entering the holding tube based on the cross-sectional area = $5.8\times10^5$ spores/s

$D_T$ = decimal reduction rate at 135 °C ($D_{135°C} = 2.15$ s)

$N$ = final population leaving the holding tube; assumed $10^{-6}$ spores/s, based on that indicated by Stumbo [5] and Pflug [4] for product sterility, considering deterioration by sporulating thermophiles as the target

So:

$F_{min\ 135°C} = [\text{Log (No) -Log (N)}]*D_T.$

$F_{min\ 135°C} = [\text{Log } (5.8\times10^5) \text{ -Log } (10^{-6})]*2.15$ s.

$F_{min\ 135°C} = 11.76*2.15$ s.

$F_{min\ 135°C} = 25.28$ s.

Therefore, for a 3-inch outer diameter holding tube, with an initial population entering the holding tube of $5.8\times10^5$ spores/s and a final population of $10^{-6}$ spores/s, a holding time of 26 s at 135 °C is required to cause 11.76 decimal reductions in the thermophilic aerobic spore population.

## 4 Notes

The reliable design and qualification of continuous thermal processes require an integrated understanding of microbial lethality, fluid dynamics, and heat transfer. Accurately characterizing food rheology and flow regime enables robust prediction of minimum residence time and ensures that all fluid elements achieve the intended thermal exposure. The systematic approach presented—encompassing microbial target selection, kinetic parameter determination, and predictive modeling—provides a scientific foundation for meeting stringent safety targets such as low PNSU levels while maintaining the nutritional and sensory integrity of the product. The combined use of validated experimental methods and computational tools strengthens regulatory compliance and process optimization, supporting the production of safe, high-quality foods in modern industrial settings.

## Acknowledgments

The authors would like to express their deepest gratitude to Professor Dr. Pilar Rodriguez de Massaguer for her invaluable teachings throughout her distinguished career as a researcher and for her outstanding leadership in guiding and inspiring numerous studies in this field. Her scientific rigor, dedication, and mentorship have been fundamental to the advancement of thermal process validation in food safety and have greatly influenced our professional and academic development.

## References


1. Doyle ME, Mazzotta AS (2000) Review of studies on the thermal resistance of Salmonellae. [Online]. Available: http://meridian.allenpress.com/jfp/article-pdf/63/6/779/1673882/0362-028x-63_6_779.pdf
2. Toledo RT, Singh RK, Kong F. Food science text series fundamentals of food process engineering. [Online]. Available: http://www.springer.com/series/5999
3. "Laws oj Turbulent Flow in Smooth Pipes. By G. Nikuradse. (Forschungsheft, No. 356.) V. D.I Verlag, Berlin. R.M. 5 (in German) (1932) J R Aeronautical Soc, 36(264):1050–1050. https://doi.org/10.1017/S036839310011524X
4. Pflug IJ (2010) Microbiology and engineering of sterilization processes. Environmental Sterilization Laboratory
5. Stumbo CR (1973) Thermobacteriology in food processing [by] C.R. *Stumbo*. Academic Press
6. Bigelow WD, Esty JR (1920) The thermal death point in relation to time of typical thermophilic organisms. J Infect Dis 27(6): 602–617. https://doi.org/10.1093/infdis/27.6.602
7. Geeraerd AH, Valdramidis VP, Van Impe JF (2005) GInaFiT, a freeware tool to assess non-log-linear microbial survivor curves. Int J Food Microbiol 102(1):95–105. https://doi.org/10.1016/j.ijfoodmicro.2004.11.038
8. Peleg M (2024) Microbial inactivation kinetics models, survival curves shapes, and the temporal distributions of the individual germs deactivation. Food Eng Rev 16(2):163–178. https://doi.org/10.1007/s12393-024-09367-5
9. Peleg M (2000) Microbial survival curves — the reality of flat 'shoulders' and absolute thermal death times. Food Res Int 33(7): 531–538. https://doi.org/10.1016/S0963-9969(00)00088-0
10. Cerf O École Nationale Vétérinaire, "Tailing of survival curves of bacteria spores A REVIEW Tailing of Survival Curves of Bacterial Spores," 1977. [Online]. Available: https://www.researchgate.net/publication/22814917
11. Singh RP, Heldman DR (2014) Introduction to food engineering. Academic Press
12. Le Maguer M, Jelen P (1986) Food engineering and process applications. Elsevier Applied Science Publishers
13. Edgerton ER, Jones VA, Warren JA (1970) Effect of process variables on the holding time in an ultrahigh-temperature steam injection system. J Dairy Sci 53(10):1353–1357. https://doi.org/10.3168/jds.S0022-0302(70)86397-4
14. Sant'Ana AS, Rosenthal A, Massaguer PR (2009) Heat resistance and the effects of continuous pasteurization on the inactivation of Byssochlamys fulva ascospores in clarified apple juice. J Appl Microbiol 107(1): 197–209. https://doi.org/10.1111/j.1365-2672.2009.04195.x
15. Awuah GB, Ramaswamy HS, Economides A (2007) Thermal processing and quality: principles and overview. Chem Eng Process Process Intensif 46(6):584–602. https://doi.org/10.1016/j.cep.2006.08.004
16. Nolasco J, De Massaguer PR (2018) Thermal degradation kinetics of sugars under typical processing conditions found in Brazilian sugar and ethanol factories. Zuckerindustrie 143(9): 520–533. https://doi.org/10.36961/si19840

# INDEX




Anderson S. Sant'Ana (ed.), *Heat Resistant Microorganisms in Food Safety and Spoilage*, Methods and Protocols in Food Science, https://doi.org/10.1007/978-1-0716-5280-0,

Index

MIX
Papier aus verantwortungsvollen Quellen
Paper from responsible sources
FSC® C105338

If you have any concerns about our products,
you can contact us on
**ProductSafety@springernature.com**

In case Publisher is established outside the EU,
the EU authorized representative is:
**Springer Nature Customer Service Center GmbH**
**Europaplatz 3, 69115 Heidelberg, Germany**

Printed by Libri Plureos GmbH
in Hamburg, Germany